Quality Assurance for Analytical Laboratories

Quality Assurance for Analytical Laboratories

Edited by

M. Parkany
*International Organization for Standardization,
Geneva*

ROYAL
SOCIETY OF
CHEMISTRY

The Proceedings of the Fifth International Symposium on the Harmonization of Internal Quality Assurance Schemes for Analytical Laboratories held in Washington, DC, USA, 22–23 July 1993.

Special Publication No. 130

ISBN 0-85186-705-7

A catalogue record for this book is available from the British Library

Published by The Royal Society of Chemistry,
Thomas Graham House, Science Park, Cambridge
CB4 4WF

Printed in Great Britain by Redwood Books, Trowbridge, Wiltshire

Foreword

by Dr. Lawrence D. Eicher, Secretary-General, ISO (International Organization for Standardization)

Emerging global markets and the corresponding development of transnational industrial sectors have intensified the need for International Standards. One vital area where the lack of such standards can lead to great frustration is that of analytical methods. Indeed, the absence of an international consensus on analytical tools can be an important obstacle to multinational collaboration in research and development.

The growing importance to industry of analytical methodology was underlined by the ISO/IEC (International Electrotechnical Commission) Advisory Board on Technological Trends (ABTT), a multinational group of industrial and technological policy leaders which was asked to advise ISO and the IEC on global trends. Looking into the fundamental changes taking place in the evolution of technological innovation, and their impact on industry, the ABTT highlighted the increasing power of analytical instrumentation, and its transformation into tools for process control in production*.

The globalization of trade and the growing importance of analytical tools have produced an increasing demand for the international standardization of the latter. Industry is already finding in the ISO 9000 series of standards for quality assurance and quality management – adopted by more than 50 countries worldwide – a framework with a practical role to play in the development, or underpinning, of International Standards on analytical methods.

Many purchasers of the services provided by analytical laboratories are stipulating that the quality system supporting these services has the backing of a certificate of conformance to an ISO 9000 standard (principally, in this context, to ISO 9002). To conform to the ISO 9000 requirements for traceability, laboratories must be able to validate the precision of analytical measurements by an unbroken and fully documented chain from these measurements upstream to their source, the SI base units.

In the selection of the most appropriate analytical method for a standard on a specific product, inter-laboratory testing of certified reference materials (CRM's) is carried out to establish the quality parameters of the method in question. CRM's are therefore important links in the chain referred to above. They also provide basic tools for assessing the competence of testing laboratories in accreditation schemes. ISO 9000 is therefore a valuable tool for producers of CRM's.

A probable scenario for future development in the field could unfold as follows:

1. Analytical methods will be accepted as International Standards on the basis of inter-laboratory tests performed on selected CRM's.

2. Based on the results of these tests, the application of the proven CRM's for the calibration of the analytical instruments will become a requirement of the particular International Standard.

3. Market pressure will require the CRM producer to obtain ISO 9000 registration, which will act as a guarantee of the unbroken metrological chain from the CRM to the SI base units.

ISO has noted an increasing number of calls for "ISO-certified" CRM's, CRM producers, and laboratories. In the present context, no certification or accreditation mechanisms are operated by ISO. Conformity assessment activities are currently carried out by third-party bodies at national level, although many of these have, or are seeking, bilateral agreements with peers in other countries to promote the mutual recognition of the certificates of conformance which they issue. ISO has encouraged this process by developing guidelines for auditors that contribute to the harmonizing of assessment and registration activities.

At the same, ISO is sensitive to calls from the market that do, in fact, articulate a desire for ISO to operate, or oversee, some form of international mechanism for the accreditation of certification bodies. Exporters and multinational companies need their certificates of conformity – whether to products or quality system standards – to be global, and unrestricted by national frontiers.

The ISO Council has commissioned the ISO Council committee on conformity assessment (ISO/CASCO) to submit a comprehensive report by May 1993 on further means to promote and achieve the mutual recognition of conformity assessment activities, including the possibility of establishing an international accreditation mechanism.

This initiative and the already existing ISO 9000 series are highly relevant to the concerns of the producers of CRM's, the laboratories that use their products, and the end-customers for whom the laboratories perform their testing and analyses. They are also support elements for the development of the International Standards on analytical methods needed by today's increasingly global technologies and emerging worldwide markets.

---

* *A vision for the future– Standards needs for emerging technologies.* An analysis and recommendations concerning standardization approaches for new technologies, together with the findings of a global survey on future standardization needs. ISO/IEC: 1990, 68 p. ISBN 92-67-10154-4.

Contents

Introduction

by Dr. Michael Parkany, Senior Technical Officer, ISO (International Organization for Standardization)

The first step in obtaining the status of "Certified in accordance with ISO 9002" for a laboratory is to make a full and detailed internal evaluation, i.e. a study of the actual situation in the laboratory in question. This can be done by a respected specialized independent organization having auditing practice in this field.

There should be already an internal quality assurance scheme based on documents relevant to the specific needs of the laboratory in question.

The subject of the fifth ISO/IUPAC/AOAC Symposium is to harmonize existing schemes at an international level.

At the present time, when public opinion is demanding accountability of laboratories carrying out analyses related to socially sensitive issues (drug-testing, blood alcohol monitoring, AIDS-testing, purity of water and air, radioactivity in the environment, acid rain, etc.), the importance of harmonizing protocols for quality assurance schemes cannot be overemphasized.

The objective of the IUPAC/ISO/AOAC Harmonization Symposium is to draft PROTOCOLS FOR THE DEVELOPMENT AND PRESENTATION OF QUALITY ASSURANCE PROCEDURES acceptable to industrial, national and international organizations. A large number of such organizations have already expressed interest.

A detailed questionnaire has been circulated for the collection of information on present practice, and the aim of this Symposium is to emphasize common elements and to reduce differences between them. Dr. Roger Wood of the United Kingdom Ministry of Agriculture, Fisheries and Food and Dr. Michael Thompson of Birkbeck College, University of London, present a discussion document for the Symposium that they prepared together with the members of the Statistics Sub-Committee of the Royal Society of Chemistry (RSC) Analytical Methods Committee and which is commented on by invited speakers. A draft Protocol will be drawn up after the Symposium and will be circulated for further comment. In due course, these comments will be evaluated at a Harmonization Workshop, where a final IUPAC Protocol will be prepared.

It is envisaged that the document will cover at least the following topics:

What is Analytical Quality Control and Analytical Quality Assurance ? (in accordance with ISO/TC 176 standards) • Basic principles • Built-in methods for detecting errors in analytical results • Monitoring of performance • Good laboratory practice (GLP) and continuous improvement of performance; the importance of feedback • Comparison – hierarchy of modern analytical methods • Simple, sensible rules to follow • Reference materials as cornerstones for intercomparison schemes (in accordance with ISO Guide 33).

This Protocol will also be submitted to ISO/TC 69, *Application of statistical methods*, for adoption as an International standard and to AOAC for the preparation AOAC Guidelines.

These proceedings contain the lectures presented at the Fifth International Symposium on the Harmonization of Internal Quality Assurance Schemes for Analytical Laboratories.

When preparing the Subject Index it became clear that a great variety of terms is used by different authors for the same concept. This is why the correct terms and their definitions concerning standardization and related activities have been annexed.

Further, for the convenience of those wishing to procure the relevant International Standards a selective list of these is also attached.

I wish to acknowledge the work of the other members of the organizing committee, namely:

Dr. A.J. Head
Dr. G. Heavner
Dr. Sj. H.H. Olrich
Prof. G. Svehla
Dr. M. Thompson
Dr. R. Wood

We extend our thanks to the authors providing their manuscripts and to Mrs. J.A. Seakins of the Royal Society of Chemistry for arranging this special publication.

This symposium is the fifth in a series sponsored by ISO, IUPAC and AOAC International, however, but the first for which the Proceedings have been published by the Royal Society of Chemistry.

Progress Towards the Establishment of an International Chemical Measurement System

Bernard King

LABORATORY OF THE GOVERNMENT CHEMIST, QUEENS ROAD,
TEDDINGTON, MIDDLESEX TW11 0LY, UK

1   THE ISSUES

Discussions involving scientists from around the world indicate that there is a perceived need to do something more to improve the quality and comparability of chemical measurements. As a first step a one day international workshop will be held to discuss the needs and explore how existing activities can be developed to meet the needs of the twenty first century.

The last few decades have seen the development of enormously powerful techniques for the analysis of complex samples in evermore detail. This has fuelled increased customer expectation both with regard to the technical specification and value for money. Data quality is under increasing scrutiny and the provision of rapid results is often also required. These trends are certain to continue and present a serious challenge to the analytical scientist and society. In addition there are clear signs that some data are not reliable and results produced in different laboratories often do not agree.

We increasingly live, however, in a world where activities in one continent can have a serious impact on people thousands of miles away. Trade, transport, health and the environment are just some areas where interdependence is increasing. Thus it is essential that analytical laboratories across the world have appropriate means of comparing and mutually agreeing their data.

2   THE TASK

We need to develop further the systems which help analysts make reliable measurements and develop an international infrastructure which promotes and

facilitates reliable measurements.   Specific tasks
include:

The construction of a conceptual model which
provides a technical and organisational framework for
the development of improved measurement comparability

The creation of improved national and
international infrastructures to provide improved
mechanisms for:

communication, cooperation and harmonisation of
standards and working practices

collaboration leading to improved support
facilities and better utilization of resources.

3   PROGRESS SO FAR

Support structures in the form of national institutes,
professional and learned bodies and international
organisations have been under constant development for
over 150 years and the pace of development is
increasing. In Europe, for example, EURACHEM was
established in 1989 to provide a focus for analytical
quality improvement in the EC and EFTA countries and
FECS WPAC has for many years provided a forum for
national chemical societies to develop international
cooperation. Similar developments are taking place in
other parts of the world and existing international
organisations such as BIPM, IUPAC, ISO, AOAC and ILAC
are expanding their activities.

Other national and regional organisations could be
substituted for the UK and European organisations and
many additional sectoral organisations could be added
to Table 1.

Table 1   Some of the key organisations

Area of Activity	UK	Europe	International
Learned and professional	RSC	FECS	IUPAC
Metrology	) VAM	) BCR	BIPM
Quality Focus	) CHEMAC	) EURACHEM	?
Accreditation	NAMAS	WELAC	ILAC
Standards	BSI	CEN	ISO

Although we have numerous organisations, possibly too many, there are gaps. Also, although international cooperation within particular generic activities, such as professional and learned affairs, is generally effective, communication <u>between</u> activities is much less well developed. In the UK and Europe the formation of quality focus groups (CHEMAC and EURACHEM) has helped improve cooperation and provide the missing links which tie together quality related matters.

4   THE ATLANTIC WORKSHOP

A workshop entitled <u>Analytical Chemistry and the 21st Century</u> and concerned with "The Development of an International Chemical Measurement System" is planned to take place in Atlanta, USA on 12 March 1993 in association with the Pittsburgh Conference. Approximately 50 senior delegates from 30 different countries and organisations will decide what more needs to be done and hopefully generate the commitment to do it. Presentations will cover:

Progress so far
High level metrology: a role for BIPM
How can chemical societies help?
A Russian Perspective
The Needs and Contributions from Industry

Organisations will have the opportunity to make short statements about what they can contribute and a substantial part of the day will be set aside for discussion of the way forward. It is anticipated that a Working Group will be formed to plan future activities.

5   THE FUTURE

It would be unwise to try to predict the outcome of the Atlanta Workshop but it will be possible to report progress at the ISO/IUPAC/AOAC symposium in Washington in July 1993.

REFERENCE

1   B King, 'The Development of an International Chemical Measurement System' (To be published in The Analyst in 1993)

Accrediting Environmental Lead Testing Laboratories

J.W. Locke and A.A. Liabastre
AMERICAN ASSOCIATION FOR LABORATORY ACCREDITATION, 656 QUINCE
ORCHARD ROAD, GAITHERSBURG, MD 20878, USA

1   INTRODUCTION

A lead laboratory accreditation program is being developed under the
regulatory authority of the U. S. Environmental Protection Agency.
Requirements of the Department of Housing and Urban Development (HUD)
and the Consumer Product Safety Commission (CPSC) must also be
considered in implementing the program. The EPA program for
assessing laboratories will be implemented by third party laboratory
accreditation organizations. Currently both the American Association
for Laboratory Accreditation (A2LA) and the American Industrial
Hygiene Association (AIHA) are being considered by the USEPA for
possible recognition as accreditors of lead testing laboratories.

The A2LA Program offers the lead testing program with the
broader scope of accreditation including organizations engaged in
other kinds of environmental assessment activities. The specifics of
the program described herein will be modified when EPA completes its
requirements.

The program is designed to accredit laboratories that conduct
assessment activities associated with determining the presence of
lead in environmental samples and the extent of this contamination.
The assessment involves field testing, sample collection, and
laboratory analysis in association with lead contamination
originating from lead-containing paint and other similar sources of
lead. The program and the attendant accreditation is available to
organizations that conduct any or all of these activities. The lead
of concern is usually found in several matrices which include air,
building debris, dust, paint residue (chips), soil, and water. The
main test technologies include:

Atomic Absorption Spectroscopy - Flame (AAS-Flame);

Atomic Absorption Spectroscopy - Furnace (AAS-Furnace);

Inductively Coupled Atomic Emission Spectroscopy (ICP-AES); and

X-Ray Florescence Spectroscopy (XRF).

The lead testing area is currently undergoing extensive research and
regulatory scrutiny which has resulted in a number of efforts to

develop methods capable of providing valid analytical procedures for
the analysis of lead contamination. A number of these methods are in
the final draft stages. This A2LA program endorses the use of these
methods as appropriate to the matrix of interest. The methods
acceptable for use under this program are listed below. These
methods will be superseded by either adoption of the respective final
version or when research or best practice indicates that a specific
method is no longer acceptable for use. The U.S. Environmental
Protection Agency (USEPA) has developed measurement protocols for
several different lead measurement methods (40 CFR: 50, 136, 141,
261; and SW 846 3rd Ed.) and has several draft methods undergoing
final development. ASTM has developed measurement protocols for
several different lead measurement methods and has a number a draft
methods under development. The National Institute of Occupational
Safety and Health (NIOSH) has developed measurement protocols for the
analysis of airborne lead and dust lead measurement methods. A2LA
provides accreditation for any of these methods.

There are strong opinions about the applicability of some of the
procedures to certain types of sample matrices or types. A2LA does
not intend to recommend which procedures are to be used in particular
situations except to require that the methods chosen (from the
acceptable methods list) be followed in detail. The application of
the method must also remain consistent with its scope. A2LA will
attest to the competence of laboratories performing to the current
state of the art.

An important aspect is the choice of the methods which are to be
used to analyze for lead in environmental samples. Method choice
will depend on a number of variables such as sample matrix,
concentration range, necessary sample preparation, detection limit,
dynamic range, precision, potential interferences, ease of use, and
cost. There are seven sources of environmental samples that may be
contaminated with lead: air, building debris, dust, paint
(unapplied), paint residue, soil, and water. The choice of methods
is limited to those methods of demonstrated performance or that are
undergoing validation/development and are currently regarded as the
best available technology and/or method. These methods are
identified in the following section of this paper.

2   PROPOSED METHODS LIST

Sample Preparation

EPA SW-846 3rd Ed. Method No. 3015: Microwave Assisted Acid
Digestion of Aqueous Samples and Extracts.

EPA SW-846 3rd Ed. Method No. 3050: Acid Digestion of
Sediments, Sludges, and Soils.

EPA SW-846 3rd Ed. Method No. 3051: Microwave Assisted Acid
Digestion of Sediments, Sludges, Soils, and Oils.

EPA SW-846 3rd Ed. Method No. 1311: Toxic Characteristic
Leaching Procedure (TCLP).

EPA CLP SOW for Inorganics Document No. ILM01.0.

EPA FR Vol.56, No.207, Friday, October 25, 1991, CEM Microwave Digestion Procedures.

EPA Method No. 200.2. Sample Preparation Procedure for Spectrochemical Determination of Total Recoverable Elements.

EPA (Draft) SOP September 1991. Standard Operating Procedure for Lead in Paint By Hotplate- or Microwave-based Acid Digestions and Atomic Absorption or Inductively Coupled Plasma Emission Spectroscopy.

NIOSH Method No. 7082: Lead in air.

NIOSH Method No. 7300: Elements in Air.

ASTM Designation: D 4309-91. Standard Practice for Sample Digestion Using Closed Vessel, Microwave Heating Technique for the Determination of Total Recoverable Metals in Water.

ASTM Designation: C 702-87. Standard Practice for Reducing Field Samples of Aggregate to Testing Size.

ASTM (Draft) Standard Test Method for the Preparation of Dust Samples, Obtained Using Wipe Sampling, for Subsequent Analysis by Flame Atomic Absorption (FAAS), Inductively Coupled Plasma (ICP-AES), or Graphite Furnace Atomic Absorption (GFAAS) Techniques.

ASTM (Draft) Standard Test Method for the Digestion of Soils for Subsequent Analysis by Flame Atomic Absorption (FAAS), Inductively Coupled Plasma (ICP-AES), or Graphite Furnace Atomic Absorption (GFAAS) Techniques.

Laboratory Sample Analysis

EPA SW-846 3rd Ed. Method No. 6010: Inductively Coupled Plasma Atomic Emission Spectroscopy.

EPA SW-846 3rd Ed. Method No. 7420: Flame Atomic Absorption Direct Aspiration.

EPA SW-846 3rd Ed. Method No. 7421: Graphite Furnace Atomic Absorption.

EPA MCAWW Method No. 239.1: Flame Atomic Absorption Direct Aspiration.

EPA MCAWW Method No.s 239.2: Graphite Furnace Atomic Absorption.

EPA Reference (40CFR50-Appendix G) Method for the Detection of Lead in Suspended Particulate Matter Collected from Ambient Air.

CLP SOW for Inorganics Document No. ILM01.0: matrix (water, wastewater and hazardous waste).

EPA Method No. 200.1. Determination of Acid Soluble Metals.

EPA Method No. 200.7. Determination of Metals and Trace Elements in Water and Wastes by Inductively Coupled Plasma-Atomic Emission Spectroscopy.

EPA Method No. 200.9. Determination of Trace Elements by Stabilized Temperature Graphite Furnace Atomic Absorption Spectroscopy.

EPA DRAFT SOP September 1991. Standard Operating Procedure for Lead in Paint By Hotplate- or Microwave-based Acid Digestions and Atomic Absorption or Inductively Coupled Plasma Emission Spectroscopy. Matrix paint chips.

EPA DRAFT Method No. 1620 1989. Lead in Water, Soil, and Sediments Using ICP-AES and/or GFAA.

NIOSH Method No. 7082: LEAD in air FAAS.

NIOSH Method No. 7300: Lead in air ICP-AES.

ASTM Designation: D 3335-85a, Standard Test Method for Low Concentrations of Lead, Cadmium, and Cobalt in Paint by Atomic Absorption Spectroscopy.

ASTM Designation: D 49-90. Standard Methods of Chemical Analysis of Red Lead.

ASTM Designation: D 1844-91. Standard Test Methods for Chemical Analysis of Basic Lead Silicochromate.

ASTM Designation: D 3280-90. Standard Methods for Analysis of White Zinc Pigments.

ASTM Designation: D 3618-91. Standard Test Method for Detection of Lead in Paint and Dried Paint Films.

ASTM Designation: D 4358-90. Standard Test Method for Lead and Chromium in Air Particulate Filter Samples of Lead Chromate Type Pigment Dusts by Atomic Absorption Spectroscopy.

ASTM Designation: D 4834-88. Standard Test Method for Detection of Lead in Paint by Direct Aspiration Atomic Absorption Spectroscopy.

ASTM (Draft) Standard Test Method for the Analysis of lead in paint.

ASTM (Draft) Standard Test Method for the Analysis of Airborne Particulate Lead.

ASTM (Draft) Standard Test Method for the Analysis of Digested Samples for Lead by Inductively Coupled Plasma (ICP-AES), Flame Atomic Absorption (FAAS), or Graphite Furnace Atomic Absorption (GFAAS) Techniques.

AOAC 5.009 (1984). Lead in Paint Using Direct Aspiration Atomic Absorption.

Field Sample Analysis

EPA (Draft) SOP September 1991. Standard Operating Procedures for Measurement of Lead in Paint Using the Scitec Map-3 X-ray Fluorescence Spectrometer.

Other XRF Spectrometer methods.

Field Test Kits.

Sample Collection Techniques

These are in addition to quality system documentation covering chain-of-custody, sampling procedures and training.

ASTM (Draft) Standard Practice for the **Field** Collection of Dust Samples Using Vacuum Sampling Methods for Subsequent Digestion and Analysis of Lead by Flame Atomic Absorption (FAAS), Inductively Coupled Plasma (ICP-AES), or Graphite Furnace Atomic Absorption (GFAAS) Techniques.

ASTM (Draft) Standard Practice for the **Field** Collection of Dust Samples Using Wipe Sampling Methods for Subsequent Digestion and Analysis of Lead by Flame Atomic Absorption (FAAS), Inductively Coupled Plasma (ICP-AES), or Graphite Furnace Atomic Absorption (GFAAS) Techniques.

HUD Paint Chip/Residue Sample Collection Procedure[*1]: Punch/Cut Method.

HUD Paint Chip/Residue Sample Collection Procedure[*]: Cut Method.

HUD Paint Chip/Residue Sample Collection Procedure[*]: Heat Removal Method.

Collection of unapplied paint samples.

Soil sampling procedure.

Paint residue sampling procedure.

Building debris sampling procedure[*].

3 GENERAL CRITERIA

A2LA uses as the basis for all of its accreditations the world recognized ISO/IEC Guide 25-1990, "General Requirements for the Competence of Calibration and Testing Laboratories". These requirements have become the standard guide throughout the world and finding a laboratory competent to meet these requirements has become the basis for mutual recognition agreements with accreditation systems in other countries. A2LA currently has in force four agreements and is negotiating with systems from additional countries. In these agreements, test data from accredited laboratories in the

---

[1]. See Attachment A at the end of the paper

United States are accepted as if the laboratories were accredited in one of these other countries as well.

Several testing technologies are available and should be selected as appropriate to the sample type and associated action level. HUD, EPA and CPSC have legislative and regulatory responsibilities which they must exercise in dealing with the problems associated with the use of paint, removal and disposal of lead based paint, paint residue, building debris and contaminated soil, and the A2LA program must take these into consideration.

Accredited organizations are permitted to advertise the fact that they are accredited. Their scope of accreditation is specific, and users are encouraged to ask to see the scope of accreditation to review those specifics before employing an accredited laboratory. The A2LA Directory includes the scope of testing for each laboratory, and users may always contact the Association for specifics of a laboratory's competence.

The general criteria for accreditation of laboratories and/or field testing organizations are contained in Part A of the A2LA green booklet entitled, General Requirements for Accreditation. These are the ISO/IEC Guide 25 Requirements. All provisions except paragraph 5.2(f) of Guide 25 apply under this program.

The general criteria for field testing activities exclude sections 7 and 14, of Guide 25. For the environmental lead program, references to the laboratory in the general requirements for accreditation shall mean laboratory and/or field testing organizations as appropriate.

To summarize the general requirements for accreditation, each organization, as appropriate to their activity, shall have:

a recognizable organization and management structure;

a documented quality system with periodic audits and reviews and quality control and quality assurance procedures appropriate for the testing technologies or sample collection procedures employed;

trained and competent personnel;

calibrated testing and measuring equipment;

test methods and/or standard operating procedures available and understood;

controlled accommodation and environment as necessary;

specimens (samples) handled carefully and chain-of-custody procedures included as necessary;

records and certificates and reports reflecting the proper conduct of the sample collection or testing;

subcontractors, outside support and services of adequate quality to meet the requirements of ISO Guide 25; and a formal <u>complaints</u> handling procedure and related documentation.

4   SPECIFIC CRITERIA

Specific criteria are an elaboration on or interpretation of the general criteria plus those additional requirements applicable to a certain field of testing, testing technology, type of test, or specific test. The numbering system for each section below corresponds to the numbering system in Guide 25. The specific criteria applicable to the Environmental Lead Program are as follows:

4. <u>Organization and management</u>. No additions to Guide 25.

5. <u>Quality system, audit and review</u>. The laboratory and/or field testing organization shall comply with the quality system provisions (section 5) of Guide 25. In addition, the organizations shall comply with the quality control (QC) procedures required by applicable federal or state environmental or public health agencies when testing specific matrices.

Standard curves shall be prepared to adequately cover the expected concentration ranges of the samples using at least 3 calibration standards and one blank, unless otherwise specified by the method employed. New curves shall be prepared whenever an out-of-control condition is indicated and after new reagents are prepared.

Field testing devices shall be calibrated as required by the testing procedure. In the absence of a requirement in the testing procedure, calibration shall be in accordance with the manufacturer's specification.

Control chart data or the equivalent shall be maintained for each routine analysis or testing activity. A documented corrective action plan shall be implemented when analytical results fail to meet QC criteria. Records shall indicate what corrective action has been taken when results fail to meet QC criteria.

Supervisory personnel shall review the data calculations and QC results. Deviations or deficiencies in QC shall be reported to management, and such reports shall be recorded. QC data shall be retrievable for all analytical and/or testing results. Method detection limits shall be determined and documented.

The laboratory shall conduct routine analyses of reagents, water used for dilutions, and solvents used for extractions to document the absence of contamination. Trip, field, and laboratory blanks shall be routinely analyzed as needed.

The laboratory and/or field testing organization shall have QC procedures (SOPs) specific to each test technology addressing, as appropriate the use of:

o   reagent/method blank analyses;

o   trip and field blanks;

o replicate/duplicate analyses;

o spiked and blank sample analysis;

o blind samples;

o surrogate standards;

o quality control samples;

o control charts;

o calibration standards and devices;

o reference material samples; and

o internal standards.

The following minimum QC sample analysis program shall be practiced in the laboratory:

o one QC check standard (instrument check solution) in 20 samples tested; the lab should repeat all samples if QC check standard is outside $\pm$ 10%;

o one blank in 20 (or per batch) both field and/or (reagent) laboratory;

o one spike in 20 (or per batch). The spike must be prepared from a standard stock which is different from the calibration standard stock, and should have a lead concentration that is within the range of the samples to be run;

o one (matrix) duplicate or (matrix) spiked duplicate in 20 (or per batch) independently prepared samples run as blinds; and

o one reference control sample (consists of a representative matrix spiked with the target analytes) in 20 (or per batch). This reference material is a secondary reference material whose concentration is traceable to a primary reference material.

Realistic sample matrices are to be used for the reference materials.

6. _Personnel_. The laboratory and/or field testing organization shall comply with all staff/personnel provisions (section 6) of Guide 25. In addition, the laboratory and/or field testing organization shall have documented evidence of analyst/tester proficiency for each test method performed. Persons in each senior technical position shall have a bachelor's degree in one of the applied sciences as a minimum educational requirement. Each analyst/tester accountable for performing tasks in any of the following areas shall meet the associated specified minimum experience requirements:

o general chemistry and instrumentation -- six months;

o atomic absorption -- one year;

o   atomic emission spectrometry -- one year;

o   x-ray fluorescence spectroscopy -- two years;

o   field testing -- six months; and

o   sample collection -- six months.

7.   <u>Accommodation and environment</u>.   The laboratory (this section does not apply to field testing) shall comply with the environment provisions (section 7) of Guide 25.   In addition, the laboratory environment shall:

o   use distilled/demineralized water that it can demonstrate to be free of interferents at detection limits;

o   routinely check and record the conductivity of distilled/demineralized water (for a continuous system check should be per batch or daily);

o   provide exhaust hoods for volatile materials [per 29 CFR (Code of Federal Regulations) 1910.1450, Occupational Exposure to Toxic Substances in Laboratories];

o   provide contamination-free work areas (as necessary);

o   provide adequate facilities for storage of samples, extracts, reagents, solvents, reference materials, and standards to preserve their identity, concentration, purity, and stability;

o   have written detailed procedures and facilities in place for collection, storage, and disposal of chemical wastes (40 CFR 261);

o   appropriately store corrosive, reactive, or explosive chemicals safely in conformance with 29 CFR 1910; and

o   provide adequate separation of activities to ensure that no activity has an adverse effect on analyses.

While specific safety criteria are not an aspect of laboratory accreditation, laboratory personnel should apply general and customary safety practices as a part of good laboratory procedures. Each laboratory must have a safety and chemical hygiene plan [per OSHA (Occupational Safety and Health) rule 29 CFR 1910], as part of their standard operating procedures.   Where safety practices are included in an approved method, they must be strictly followed.

8.   <u>Equipment and reference materials</u>.   The laboratory and field testing organization shall comply with the equipment and reference materials provisions (section 8) of Guide 25.   Equipment used for lead based paint testing shall meet the following minimums:

<u>For analytical balances/pan balances</u>:

o   analytical balances shall be capable of weighing to 0.1 mg.;

o records of balance calibration shall be kept for at least two ranges (no more than two decades apart) using weights that conform to at least Class 3 tolerances (ASTM E 617-1990);

o records showing daily functional/calibration checks for analytical balances and monthly for other balances shall be maintained; and

o the balances shall undergo metrological calibration at least annually.

For pH meters:

o the laboratory shall use a clean pH meter with properly maintained electrodes suitable for the test performed with scale graduations at least 0.1 pH units (calibrated to $\pm$ 0.1 pH units for each use period);

o either a thermometer or a temperature sensor for temperature compensation shall be in use. Automatic temperature compensators which are an integral part of the apparatus are acceptable.

o a magnetic stirrer with clean PTFE-coated spin bar shall be utilized when making pH measurements;

o records shall be kept showing daily, or before each use, calibration, whichever is less frequent. Verify the absence of electronic drift by analyzing a reference buffer as a sample at least every 20th sample or at least once per batch; and

o aliquots of standard pH 4 & pH 7, or pH 7 & pH 10 shall be used only once.

For labware and sample collection devices:

o all such devices shall be cleaned in a manner appropriate for the analytical procedures for which they are to be used.

For ovens:

o thermometers shall be graduated in increments no larger than 1°C;

o if oven temperature cannot be read without opening the door, the bulb of the thermometer shall be immersed in a sand bath; and

o oven temperature shall be adequately monitored (e.g., beginning and end of each use cycle).

For hot plates:

o maintain temperature at the center of the hot plate such that the surface temperature is 140°C. Note: An uncovered beaker containing 50 ml of a liquid such as an oil (water?) located in the center of the hot plate can be maintained at a temperature no higher than 140°C.

For microwave ovens:

o calibrate the power available for heating weekly. This
quality control function is performed to determine that the microwave
has not started to degrade and that absolute power settings (watts)
may be compared from one microwave unit to another. This power
evaluation is accomplished by measuring the temperature rise in 1 kg
(1.0 liter) of water exposed to microwave radiation for a fixed
period of time. Water is placed in a teflon beaker and stirred
before measuring the temperature. The beaker is circulated
continuously through the field for 2 minutes with the unit at full
power. The beaker is removed, the water vigorously stirred, and the
final temperature recorded. The final reading is the maximum
temperature reading after the energy exposure. These measurements
should be accurate to $\pm$ 0.1°C and made within 30 sec of the end of
heating. The absorbed power is determined by the following
relationship:

$$P = \frac{(K)\ (CP)\ (m)\ (T)}{t}\ ;$$

Where:

P = the apparent power absorbed by the sample in watts (W),
[W=joule per sec].

K = the conversion factor for thermal capacity or specific heat
(cal per gm per degree C) of water.

Cp = the heat capacity, thermal capacity, or specific heat (cal
per gm per degree C) of water.

m = the mass of the water sample in grams.

T = T_f, the final temperature minus the T_i, the initial
temperature in degrees C.

t = time in seconds (s).

Using 2 minutes and 1 Kg of distilled water, the calibration
equation simplifies to: P = (T) (34.87). The power in Watts can now
be related to the percent power setting of the microwave unit. The
microwave is then calibrated by simply plotting the percent power
rating versus the experimentally determined Watts.

For thermometers:

o the laboratory shall have access to a NIST (NBS)-traceable
thermometer for use in verifying working thermometers;

o the calibration of working mercury-in-glass thermometers
shall be checked at least annually against a NIST (NBS)-traceable
certified thermometer; and

o the calibration of dial-type thermometers shall be checked at
least quarterly against a NIST(NBS)-traceable thermometer.

For autopipetors/dilutors:

o the apparatus shall have sufficient sensitivity for the intended use; and

o records shall be kept showing delivery volumes are checked gravimetrically at least monthly.

9. Measurement traceability and calibration. The laboratory and\or field testing organization shall comply with the measurement traceability and calibration provisions (section 9) of the general criteria. In addition, the organizations shall as appropriate:

o document the frequency, conditions, and standards used to establish calibration all analytical/testing methodology; and

o verify and document all working standards versus primary (reference) standards.

10. Calibration and test methods. The laboratory and/or testing organization shall comply with the calibration and test method provisions (section 10) of Guide 25. In addition, the organizations shall:

o use approved (EPA, HUD, ASTM, NIOSH accepted and/or draft methods as appropriate) methodologies in their entirety as required for each test or analysis performed;

o have procedures for making and controlling revisions to in-house SOPs (use revised SOPs only after written authorization by senior technical personnel);

o have documented procedures for data collecting and reducing, reporting and record keeping;

o have documented validation procedures to apply at appropriate levels of all measurement processes;

o have documented procedures to check the validity of reported analysis values;

o have documented procedures for correcting erroneously reported results;

o have quality control procedures documented and available to the analysts/testers;

o use reagent grade or higher purity chemicals to prepare standards;

o use primary standard & QC reference materials;

o prepare fresh analytical standards at a frequency consistent with good laboratory practices unless otherwise stated in the method (frequency is a function of concentration and type of matrix); generally, the lower the concentration the less stable the standard) o properly label reference materials/reagents with concentrations, date of preparation, expiration date and the identity of the person preparing the reagent; and

o have standards preparation documentation such as a preparations record book.

11. <u>Handling of calibration and test items</u>. The laboratory and/or testing organization shall comply with the handling provisions (section 11) of Guide 25. In addition, the organizations shall:

o have documented procedures for collection, shipping, receipt and storage of samples as appropriate.

o give samples an unambiguous sample number when collected and/or logged.

o maintain a permanent record for sample collection and log-in data;

o store samples in such a way as to maintain their identity, integrity, stability, and concentration; and

o follow documented chain-of-custody procedures, when required.

The organization shall have a sample custodian who shall be responsible for the sample control/logging. The procedures involved include the control, identity, preservation, and condition of samples, and sample handling, storage, and disbursement for analysis. The laboratory shall have a person responsible for ensuring that all analyses are performed within any USEPA/HUD or method-specified holding times, where appropriate.

12. <u>Records</u>. Test records shall be protected from loss, damage, misuse or deterioration and shall be retained for an appropriate period in a manner that permits retrieval when required. Test records that are created and/or retained on magnetic media (e.g., computer disks) or photographic media (e.g., microfiche) shall be stored in a manner that protects them from the hazards that affect such media and provision shall be made for the printing of such records when required. Note: It is not possible to define a particular retention period that is suitable for all laboratories' and/or field testing organizations circumstances. The minimum appropriate period will be based upon the nature of the organization's work, and its regulatory, legal, and contractual obligations. The organization shall have:

o a system that provides for retrievability and traceability of the sample source, the methodology of analysis/testing, results (including calibration and instrument checks), the person performing the analysis, and the date; and

o a secure archive area where records are held for appropriate periods of time and where access, deposit and removal of records are controlled and documented.

The organization shall comply with all the records provisions (section 12) of Guide 25. In addition, the organization shall establish and maintain a records system ensuring that:

o all observations and calculations are recorded in a permanent manner (such as laboratory/field notebooks, pro-forma work sheets, or magnetic media) at the time they are made and that the units of measurement in which observations are recorded are stated;

o original records are uniquely identified and traceable to the tests or test items to which they refer and to any test reports based upon them;

o records are traceable, retrievable, and legible and include sufficient information and explanation such that they can be readily interpreted by staff other than those responsible for their generation;

o records contain sufficient information to permit identification of possible sources of error and to permit, where feasible and necessary, satisfactory repetition of the test under the original conditions;

o records contain sufficient details of any significant departures from test specifications or other specified procedures including authorizations for such departures;

o records are checked for data transcription or calculation errors;

o records identify the person or persons responsible for their generation and those responsible for checking data transcriptions and calculations; and

o corrections or amendments to test records are made in a manner that does not obliterate the original data and are signed or initialled by the person responsible.

13. Certificates and reports. Test reports shall include a signature of the analyst/tester who conducted the test and shall conform to the documentation requirements listed in attachment B.

14. Sub-contracting of calibration or testing. No additions to Guide 25.

15. Outside support and supplies. No additions to Guide 25.

16. Complaints. No additions to Guide 25.

5 PROFICIENCY TESTING [PERFORMANCE EVALUATION (PE)] REQUIREMENTS

For potable and nonpotable waters, the water supply (WS) and water pollution (WP) programs are the performance evaluation (PE) studies offered by EPA. To be accredited for testing in these matricies, each participating laboratory must supply A2LA headquarters with a copy of PE results as soon as they are received at the laboratory.

Every laboratory getting PE results that are other than "acceptable" as determined by EPA for a specific analyte (contaminant) must investigate and identify likely causes of other than acceptable results, resolve any problems and report such activity to A2LA headquarters along with the submittal of PE results.

If one result out of a pair of two concentrations is found to be other than acceptable by EPA, the laboratory must provide an explanation of why it believes such results were obtained. If a laboratory obtains other than acceptable for both results in a pair, the laboratory must provide an explanation and obtain suitable quality control samples on its own, perform the requisite tests, and report the results of these tests, indicating that it is obtaining appropriate results. Proficiency tests from other state programs may provide suitable evidence that a laboratory can obtain acceptable results.

If a laboratory fails to provide sufficient evidence that it can obtain acceptable results and if it misses (i.e., obtains an other than "acceptable" result) on at least one concentration of the same analyte in two consecutive PE studies, the laboratory's accreditation shall be subject to revocation for that analyte unless it agrees to seek help from an outside technical expert suitable to A2LA. The laboratory shall be responsible for all technical expert costs.

If the laboratory then does not obtain acceptable results (including both concentrations for sample pairs) during the next (third consecutive) PE study for the analyte(s) in question, the laboratory will be subject to automatic revocation for that analyte.

To be accredited for samples originating from the lead hazard assessment and abatement activities associated with minimizing and eliminating childhood lead poisoning, required proficiency testing program is the performance evaluation (PE) studies offered through AIHA for NIOSH/EPA/HUD known as the Environmental Lead Proficiency Analytical Testing (ELPAT) Program. Samples will be supplied on a quarterly basis. Each participating laboratory must supply A2LA headquarters with a copy of PE results as soon as they are received at the laboratory. Every laboratory getting PE results that are other than "acceptable" as determined by NIOSH for lead in a specific matrix must investigate and identify likely causes of other than acceptable results, resolve any problems and report such activity to A2LA headquarters along with the submittal of PE results. Four concentration levels will be required for each of three matrices: Paint chips, soil, and dust wipes.

If the result for a particular matrix is found to be other than acceptable by NIOSH, the laboratory must provide an explanation of why it believes such results were obtained. In addition to providing an explanation the laboratory must provide obtain suitable quality control samples on its own, perform the requisite tests, and report the results of these tests, indicating that it is obtaining appropriate results. Proficiency tests from other programs may provide suitable evidence that a laboratory can obtain acceptable results.

If a laboratory fails to provide sufficient evidence that it can obtain acceptable results and if it misses (i.e., obtains an other than "acceptable" result) on the same matrix in two successive PE studies, the laboratory's accreditation shall be subject to revocation for that matrix unless it agrees to seek help from an outside technical expert suitable to A2LA. The laboratory shall be responsible for all technical expert costs.

If the laboratory then does not obtain acceptable results during the next (third consecutive) PE study for the matrix in question, the laboratory will be subject to automatic revocation for that matrix.

The field testing and sampling organizations shall be required to participate in suitable proficiency testing programs as they become available. The listing of any accredited laboratory will not be continued in the yearly A2LA Directory of Accredited Laboratories unless all relevant proficiency test data requirements have been met.

ATTACHMENT A: HUD PAINT CHIP/RESIDUE COLLECTION PROCEDURES.

The sampling procedures described here are based on guidance provided in the Office of Public and Indian Housing, Department of Housing and Urban Development, "Lead-Based Paint: Interim Guidelines for Hazard Identification and Abatement in Public and Indian Housing," September 1990.

There are several sampling situations that are presented when collecting paint chip samples. The paint should be removed in a way that minimizes the adherence of substrate to the paint film (paint chips). All layers of paint down to the substrate are to be included. Samples taken for laboratory analysis should be about 2 square inches in size. Three general methods for sample collection follow:

a.   Punch/cut Method:

Collect a sample of known area. The results of analysis for samples collected using this method are to be reported in the units of mass per unit area, mg/cm^2). Perform the following:

1) apply a clear adhesive material over a larger area than the sample size desired;

2) cut through the adhesive and paint layers using a punch or template and sharp knife;

3) place the sample in a sealable plastic bag and identify it with a unique sample number;

4) remove the adhesive, paint and a thin layer of the substrate using a sharp chisel; and

5) use a brush or mini-vacuum to clean the area and properly dispose of any wastes in a plastic bag.

b.   Cut Method:

Collect a sample of suitable weight. The results of analysis for samples collected using this method are to reported in the units of weight percent. Perform the following:

1) using a sharp tool, score the area of paint to be removed; then slide a thin sharp blade along the score and under the paint, removing the section of paint down to, but not including, the substrate ensuring all layers of paint are intact (if the results of analysis are to be reported in the units of weight percent then no substrate can be included); and

2) place the sample in a sealable plastic bag and identify it with a unique sample number;

3) use a brush or mini-vacuum to clean the area and properly dispose of any wastes in a plastic bag.

c. Heat Removal Method: Samples will be obtained with and without substrate, on some substrates (works well on wood and steel); with practice paint film can be removed without the substrate adhering to it. Perform the following:

1) using a heat gun, direct hot air from a distance of 4" to 6" from the surface, onto the paint while pressing the edge of a sharp tool into the softened paint [heat gently to avoid overheating and/or causing smoking (the lead sublimes)];

2) heat and cool alternately for a few seconds at a time while gently pressing the knife edge into the paint;

3) using the knife lift the paint off the substrate, then if necessary scrape the surface to remove residual paint;

4) place the sample in a sealable plastic bag and identify it with a unique sample number; and

5) use a brush or mini-vacuum to clean the area and properly dispose of any wastes in a plastic bag.

ATTACHMENT B: DOCUMENTATION REQUIRED FOR LABORATORY RESULTS

Based on HUD requirements the following documentation requirements must be met. [Lead-Based Paint: Interim Guidelines for Hazard Identification and Abatement in Public and Indian Housing, Department of Housing and Urban Development, September 1990].

The laboratory must record and report data such that the data package can be validated and legally defensible. The information and data required can be categorized or classified as listed in the following section. Each classification listed also includes the minimum data elements required. In addition, each classification may be regarded as a separate data form to be completed by the laboratory in its documentation of the analysis. The list for each classification is not exhaustive and is intended to serve as minimum guidance to the laboratory regarding the information it is responsible for collecting, documenting and reporting.

1. Cover Page Information. General information about (Identification of the laboratory) the laboratory, identification of the sample preparation and analytical method employed and the conditions under which these methods were performed are to be

presented on the cover page. The information provided must minimally include the following:

    a. laboratory identification;
    b. analytical run identification (sample numbers);
    c. date of report preparation;
    d. type of measurement detection system used (ICP, GFAA, FAA,etc.);
    e. type of instrument used;
    f. identification of the sample (digestion) procedure used;
    g. identification of the analytical method used;
    h. the signature of the authorized laboratory signatories; and
    i. general comments and observations about the sample, sample preparation and/or analytical run as appropriate.

    2. Sample Information. A list of all samples analyzed in an analysis run (both field samples and QC samples), in the order in which they were analyzed is to be included in this section. Information on the particular sample run is also included. The chronological ordering of the samples allows: QC samples to be matched with the field samples; measurement system control to be evaluated; the data set to be validated; and the demonstration of legal defensibility.
The information provided must minimally include the following:

    a. sample identification;
    b. batch identification;
    c. sample type (paint chips, dust, air, etc.);
    d. sample weight (in grams)/area (in cm^2);
    e. matrix information;
    f. dilution factors;
    g. date (samples) collected and received in the laboratory;
    h. date and time analyzed; and
    i. operator information;

    3. Results of Initial Precision and Accuracy Determination.
Results of the initial analysis runs of 4 spiked aliquots of reagent water. The information provided must minimally include the following:

    a. date and time of analysis;
    b. sample identification for the 4 aliquots;
    c. spiked concentrations;
    d. percent recoveries;
    e. mean, standard deviation, and relative standard deviation of the percent recoveries among the 4 aliquots;
    f. flag for problem detected by this analysis; and
    g. corrective actions, if any.

    4. Results of Calibration. Results of each determination of the calibration curve are summarized on this form:

    a. date and time of calibration;
    b. identification of the standards;
    c. concentration of the standards;
    d. detection limits;
    e. instrument response;

f.   slope and intercept terms of the fitted calibration curve;
and
g.   mean-square error and correlation coefficient of the fitted
calibration curve.

5.   Results of Analysis on Blanks.   Results of the analysis of
the blank samples listed on the Sample Information form are
summarized and minimally contain the following information:

a.   date and time of analysis;
b.   sample identification;
c.   type of blank (ICV, CCV, Preparation blank, Field blank);
d.   detection limits;
e.   instrument response;
f.   estimated concentration based on the calibration curve;
g.   flag for problem detected by this analysis; and
h.   corrective actions, if any.

6.   Results of Calibration Verification.   Results of the
verification procedure on the calibration curve through verification
standard samples are summarized and minimally contain the following:

a.   date and time of analysis;
b.   identification of the calibration curve ;
c.   sample identification of the calibration check samples;
d.   sample type (ICV, CCV);
e.   true concentration level of lead in the ICV and CCV;
f.   instrument response;
g.   estimated concentration based on the calibration curve;
h.   percent difference between the true and estimated
concentration;
i.   flag for problem detected by this analysis; and
j.   corrective actions, if any.

7.   Results of Tests for Accuracy.   Results of the analysis on
spike samples, laboratory control samples, and linear range analyses
are summarized and minimally include the following:

a.   date and time of analysis;
b.   sample identification;
c.   sample type (regular spike, ICS, LCS, LRA);
d.   spiking concentration;
e.   instrument response;
f.   estimated concentration of the sample;
g.   percent recovery;
h.   flag for problem detected by this analysis; and
i.   corrective actions, if any.

8.   Results of Tests for Precision.   Results of the analysis on
split and duplicate samples are summarized and should minimally
contain the following:

a.   date and time of analysis;
b.   sample identification;
c.   type of sample (duplicate spike samples, split samples, etc.);
d.   number of duplicates;
e.   spiking concentration (if any);
f.   instrument response;

    g.   estimated concentration of the sample;
    h.   relative percent difference for duplicate spike samples;
    i.   relative standard deviation for split and duplicate samples;
    j.   flag for problem detected by this analysis; and
    k.   corrective action, if any.

In addition, laboratories should document the data results from methods used to prevent and adjust for interferences and bias such as serial dilution methods and the Method of Standard Additions. Sources of standards used in the analysis should also be completely documented.

The Use of Statistics in Developing Intra-Laboratory Method Validation Guidelines

S.M. Anderson and J. Ngeh-Ngwainbi
KELLOGG COMPANY, SCIENCE AND TECHNOLOGY CENTER, 235 PORTER STREET, BATTLE CREEK, MI 49017, USA

1 INTRODUCTION: MEASUREMENT AS A PROCESS

A "process" has three characteristics which identify it: Sequence, Repetition and Variability. Sequence means that some functions or operations must be completed before another can begin. Repetition means that the process can be repeated as often as necessary, but with each repetition the result of the process will vary from the earlier and the later results. The variability occurs because all physical measurement is subject to error. These characteristics describe the measurement process known as analytical chemistry. The use of appropriate statistical techniques will help an investigator minimize the variability of analytical results.

Measurement error in analytical chemistry is of two types: determinate (systemic or special cause) errors and indeterminate (random) errors. Determinate errors are associated with a definite cause, even though the cause may not be readily identifiable by the investigator. They may be unidirectional (always positive or negative) with magnitude constant from sample to sample or proportional to sample size. Possible sources of determinate errors include: instrumental defects, reagent impurities, operator and/or method error. Indeterminate or random errors are due to non-permanent causes and include instrumental and environmental noise. Instrumental noise is a composite of many factors which may be difficult to characterize and is often intrinsic to system electronics. Random errors follow the familiar bell-shaped distribution known as Gaussian or "normal" whose properties are well known.

In addition to the normal distribution, some statistical techniques which prove useful in analytical chemistry are calculation of the sample mean and median, the range, standard deviation, coefficient of variation, and confidence intervals for mean.

2 STANDARDS, REFERENCE MATERIALS, AND CONTROL SAMPLES

If possible, the first step in method development (validation) is to obtain a standard material whose character and purity is well studied. It should remain stable under appropriate conditions with certifiable analyte values. These materials do not have to go through sample preparation and are used to calibrate or standardize instruments.

In lieu of a well studied standard material, other reference materials can be used to perform the same functions. These materials are, however, generally not pure or well characterized; but their concentrations are known. Most often they must undergo sample preparation and are used to calibrate the method itself or the instrument to be used.

Control samples are still a third reference material. They are products with matrices similar to samples for which the method is being developed. Their concentrations of analyte are near the middle of a linear range and are as close to typical samples as possible. They are homogeneous for the analyte in question, they are easily available, and have similar behavior to unknown samples in extracting solutions. Control samples will be used, as their name implies, in conjunction with control charts after method development is complete.

3 LINEAR CALIBRATION CURVES

The response (output) of a method can be measured as a function of increasing concentration of the analyte. At the low end of concentration, response is close to the noise of the analytical instrument. At the upper end of concentration, response flattens out, even as concentration continues to increase. The interval between the low and high concentrations is considered to be the linear response range. Within this range the best linear calibration curve is calculated using a statistics procedure known as "ordinary least squares" or "single variable linear regression". This procedure permits the following estimates: The slope and intercept of the curve; the correlation coefficient (r) which measures the amount of association between concentration and response; the adjusted R-squared value which indicates how much variation in responses can be explained solely by changing concentrations; and the calculation of confidence intervals around predicted response values for given levels of concentrations. Finally, the difference between observed responses and predicted responses, known as residuals, can be obtained using linear regression.

4 LIMIT OF DETECTION

Investigation with an appropriate material near the lower end of the concentration scale will enable an investigator to determine a lower detection limit for the method. If a matrix blank is available, the limit of detection can be calculated as the average of the matrix blank plus 3 standard deviations of the matrix blank.

5 RECOVERY STUDIES

Recovery studies are one way to evaluate the accuracy of a method. Using a standard at known concentrations, one may compare the response from a prepared sample to the response obtained using the standard material. As we have already suggested, all responses from a given method will have variation. This occurs from random error, if nothing else. Consequently, the important question in recovery studies is: "How close are the expected results from a standard sample compared with the results from a prepared sample?" This question is properly answered with the use of the statistics procedure known as "Student's t-test techniques" applied for each level of concentration.

6 METHOD VALIDATION

In general, validation refers to a process by which one verifies that a procedure does what it is supposed to do. For analytical methods, the aim of validation is to reliably and accurately measure the quantity of an analyte in a sample. When performing a method validation it is desirable to cover the entire working linear range which was determined in a linear calibration study described above.

A validated analytical method must be rugged enough to withstand the minor variations which are inevitable in the day to day operation of an analytical lab; such as, substitute lab personnel, different instruments, fresh batches of reagents, etc. A statistical technique attributed to W.J. Youden called "ruggedness testing" allows investigation of up to seven of these sources of variability at one time. For this technique, it is assumed that most of the sources of variability will have only minor effects on the method. When this hypothesis is tested according to a particular designed experiment and if results indicate no effect due to the sources, a method can be described as rugged, and "therefore immune to modest (and inevitable) departures from some habitual routine..."[1]. Traditional analysis of variance techniques can also be used, but generally result in more samples being subjected to the analytical procedure.

7 METHOD COMPARISON

A new or improved analytical method which has met all validation criteria is now ready to be compared to an existing official method whose performance is scientifically documented. Statistical techniques which allow comparisons of means and variances are useful in these instances. Also, since method comparisons usually cover the assay's linear range, simple linear regression techniques are helpful is ascertaining a new or improved method's suitability.

8 CONTROL CHARTS

The use of control charts as a statistically based decision support tool to help identify and control variation in an analytical process has been widely recognized for many years. Both qualitative and quantitative data can be control charted by an analyst, and within each of these two categories there exist many options for the appropriate kind of presentation and analysis of their data.

9 SUMMARY

What is presented here is meant to be a guideline for the use of statistics in method development and improvement. When used in conjunction with experienced analytical technicians and lots of common sense, the resulting methods should stand the test of time.

---

[1]Youden, W.J. "Statistical Techniques for Collaborative Tests." Statistical Manual of the A.O.A.C. Association of Official Analytical Chemists, Washington, D.C.

The Improvement of Quality of Observed Values – Justification for Using Intermediate Precision Conditions (ISO 5725 Part 3)

Takashi Miyazu and Hisashi Yamamoto

DEPARTMENT OF MANAGEMENT ENGINEERING, THE NISHI-TOKYO
UNIVERSITY, UENOHARA-MACHI YAMANASHI-KEN, JAPAN 409-01

0   Introduction

The first step in statistical process control and/or research studies should be the improvement of the quality of observed values because, the less reliable the data, the less reliable the results. In other words, " Garbage in, garbage out (GIGO)". In most cases, the most important quality of observed values used in the above purposes is "accuracy".

The purpose of this paper is to explain ways of improving the accuracy of observed values.

1   Definitions of terms relating to "accuracy"

The term of "accuracy" of observed values has been inconsistent used in the past. For example, in the former BS in the 1940's, it was the same as "precision"; in the former ISO 5725 (the Determination of the repeatability and Reproducibility) in 1981, it was defined as "the degree of bias". However, in the revised ISO 5725 (1993 forthcoming), the term "accuracy" and related ones are defined as follows:

4.1.6 accuracy: The closeness of agreement between the test result and the accepted reference value.

NOTE – The term accuracy, when applied to a set of observed values described a combination of random components and a common systematic error or bias component.

4.1.7 trueness: The closeness of agreement between the average value obtained from a large series of test results and an accepted reference value.

NOTE – The measure of trueness is usually expressed in terms of bias.

4.1.8 bias: The difference between the expectation of the test results and an accepted reference value.

NOTE – Bias is a systematic error in contrast to random error. There may be one or more systematic error components contributing to the bias.

A larger systematic difference from the accepted reference value is reflected by a large bias value.

4.1.9 precision: The closeness of agreement between independent test results obtained under prescribed conditions.

NOTES 1 Precision depends only on the distribution of random errors and does not relate to the accepted reference value.

2 The measure of precision is usually expressed in terms of imprecision and computed as a standard deviation of the test results. Higher imprecision is reflected by a larger standard deviation.

3 "Independent test results" means results obtained in a manner not influenced by any previous result on the same or similar material.

2 Ways of improving the accuracy of observed values

The following ways should be effective in improving the accuracy of observed values:

a) The employment of a suitable sampling and sample preparation procedure. This is very important, especially for bulk materials, such as ores and coals, because they are generally not homogeneous. In other words, a purposive sampling may easily result with such heterogeneous materials.

b) The technological development and/or improvement of the measurement methods themselves.

This is not dealt with here because it is beyond the scope of this paper.

c) The employment of the "Intermediate Precision Conditions" in the case of repeating the measurements on the same sample.

In order to improve the "trueness" of mean values, the measurements on the same sample are usually replicated. However, the former ISO 5725-1981 specified only the repeatability condition (i.e. when all factors - the time, operator and equipment - were the same).
In comparison with the intermediate precision conditions (as specified in the revised ISO 5725), the repeatability condition results in the highest precision but the lowest trueness of the observed values. In other words, the trueness of the observed values may be improved by changing only the replication condition. This forms the main focus of this paper.

d) The checking and exclusion of outliers in the observed values.

This is another effective way of improving the accuracy of the observed values. This procedure should be carried out both technically and statistically.
This is not dealt with in this paper because the statistical outlier tests are described in precise detail in the revised ISO 5725.

3 Intermediate Precision Measures (I.P.M.)

3.1 What I.P.M. are

I.P.M. are the precision measures obtained under intermediate precision conditions. The M-factor different intermediate precision conditions (M=1,2 or 3) can be defined as:
  - a) M=1 - only one of the three factors (operator, equipment, or time) is different, or the same equipment is recalibrated, between successive determinations.
  - b) M=2 - two of the three factors are different.
  - c) M=3 - all three factors differ between successive determinations.

The concept of the relation between repeatability, I.P.M. and Reproducibility concerning the values of standard deviation is shown in Fig.1.

One of the actual examples obtained by international Round-Robin studies which appeared in ISO/DIS 5725-3(1991) is reproduced in Fig.2.

3.2 The reasons for specifying I.P.M. in the revised ISO 5725

I.P.M. can be justified by the following two purposes:

1) Statistical process control with two or three shifts for which precision is more important than trueness.

In this case, precision should be obtained by intermediate precision conditions because the operator and the time necessarily change with the different shifts.

As is evident from formulae (1) and (2), precision due to the measurement, s_M, in formula (2), should not be repeatability but I.P.M..

$$S_T^2 = S_M^2 + S_P^2 \quad -------------- \quad (1)$$
$$\therefore \quad S_P^2 = S_T^2 - S_M^2 \quad -------------- \quad (2)$$

where, S_T^2; total variance observed between shifts
  S_M^2; variance due to the measurement
  S_P^2; variance due to the process variation between shifts

If we use repeatability instead of I.P.M., we may overestimate the process variation, because repeatability (s_r) is substantially smaller than I.P.M. (s_I) as shown in Table 1 and Fig.1.

2) Measurements used in international trading for which trueness is more important than precision.

For example, the estimate of the true value of a consignment is very important because it corresponds to the price of the consignment. On the other hand, there is usually no relation between repeatability and trueness (see Fig.3). However, if we employ I.P.M. conditions instead of the repeatability condition, a "trade-off relation" can be observed as shown in Table 1.

The reason can be explained as follows:

$$y_{ij} = \mu + B_i + e_{ij} \quad ----------------------------------- \quad (3)$$
$$V(y_{ij} \mid \text{repeatability condition}) = V(e_{ij}) \quad -------------- \quad (4)$$
$$V(y_{ij} \mid \text{I.P.M. conditions(M=3)}) = V(B_i) + V(e_{ij}) \quad ---- \quad (5)$$

thus, $\quad$ E ($\bar{y}$ | repeatability condition) $\quad = \mu + B_i$ ------------ (6)

$\quad\quad$ E ($\bar{y}$ | I.P.M. conditions(M=3))= μ -------------------- (7)

where $\quad$ y$_{ij}$: jth observed value at ith laboratory

$\quad\quad$ μ $\quad$: true value

$\quad\quad$ B$_i$: laboratory component of bias

$\quad\quad$ e$_{ij}$: random error occurring in every result

In other words, under repeatability conditions, the laboratory component of bias B$_i$ remains constant in the observed values. Under intermediate precision conditions (M=3), fixed effect B$_i$ varies to random effect by changing all factors (the operator, the time, the equipment). As a result, E(y$_{ij}$) under I.P.M. conditions should have less bias, in comparison with that under the repeatability condition.

3.3 How to obtain I.P.M.

Parts 1 and 2 of the revised ISO 5725 describe the precise manner for obtaining repeatability and/or Reproducibility by Round-Robin studies. I.P.M. can be obtained by just changing the repeatability conditions mentioned in Parts 1 and 2 to the intermediate conditions mentioned in Part 3.

If we wish to obtain repeatability and I.P.M. in the same Round-Robin study, a special design of experiment called "Staggered Nested Experiment" is useful. It is shown in Annex B of Part 3 in the revised ISO 5725.

4 Conclusion

As explained in this paper, the quality of observed values, i.e. accuracy, is a crucial factor which greatly affects the success or failure of process control.

This is why the authors have described in detail the improvement of the quality of observed values. Of the four effective ways outlined in Section 2, the "intermediate precision measures" is the least well-known as it is the most recent, appearing in ISO/DIS 5725 Part 3 (1991) for the first time, and it was decided to be published as ISO standard 5725 Part 3 in 1993, at the London meeting of ISO/TC69(1992).

However, I.P.M. was, in fact, specified in JIS Z 8402 (General rules for permissible tolerance of chemical analysis and physical tests) as long ago as 1974 and was already being used in Japan, mainly for industrial purposes, such as process control.

It is the sincere hope of the author that the significance of I.P.M. will be recognised and applied by many countries in the near future because of its important contribution to successful process control and/or research studies.

Acknowledgement

The authors wish to thank Ms. Ann Jenkins, lecturer of The Nishi- Tokyo University, for her help in the writing of this paper.

References

· T. Miyazu, H. Yamamoto "Intermediate measures of precision of a test method"
                      Workshop Accuracy 91 (Geneva, 1991)
· ISO/DIS 5725 Accuracy (trueness and precision) of a test method and results
  (1991)

Table 1   Relations between precision and trueness

Measurement conditions	Precision s.d.	Trueness $E(\bar{x}) - \mu = \delta$
repeatability time, operator, equipment all factors – same	s_r	δ_r
Intermediate Precision Measures One – factor – different Two – factors – different Three – factors – different	$\wedge$ $s_{I(1)}$ $\wedge$ $s_{I(2)}$ $\wedge$ $s_{I(3)}$	$\wedge$ $\delta_{I(1)}$ $\wedge$ $\delta_{I(2)}$ $\wedge$ $\delta_{I(3)}$
Reproducibility all factors – different	$\wedge$ s_R	$\wedge$ δ_R

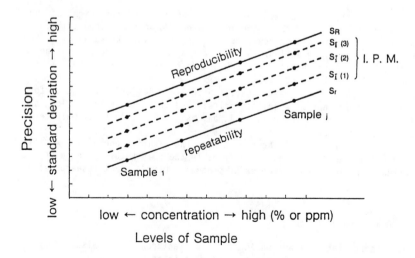

Fig.1   The Concept of the relation among three
           types of precision

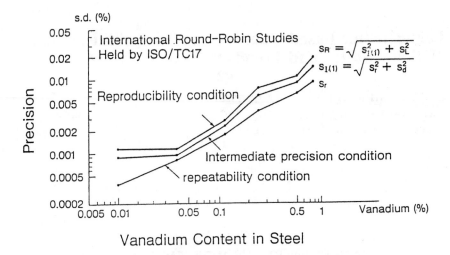

Vanadium Content in Steel

Fig.2   Relation Between Sample Concentrations
and Three Types of Precision

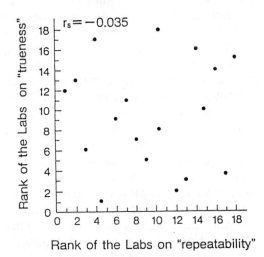

Rank of the Labs on "repeatability"

Fig.3   An Example of the Round–Robin Study
Determination of Sulphur in Heavy Oil.

Performance Based Quality Assurance of the NOAA National Status and Trends Project

A.Y. Cantillo and G.G. Lauenstein
NATIONAL OCEANIC AND ATMOSPHERIC ADMINISTRATION, NOS/ORCA
N/ORCA 21, ROCKVILLE, MD 20852, USA

1   NOAA NATIONAL STATUS AND TRENDS PROGRAM

Long-term monitoring studies, whether local, regional, or global, require that data of known quality be generated by all participants. These data must be comparable to one another in space and time, and be traceable to a common reference point. Since 1984, the National Oceanic and Atmospheric Administration (NOAA) has conducted the National Status and Trends (NS&T) Program, a long-term monitoring study of the coastal waters of the United States. Since its inception, quality assurance has played a major role in the design of the program, the evaluation of potential analytical contractor laboratories, and the maintenance of data quality.

NOAA's NS&T Program determines the current status of, and any changes over time, in environmental health relative to toxic contaminants in estuarine and coastal waters of the United States. Sampling sites are located on the East, West and Gulf coasts, Alaska, Hawaii, the Great Lakes and Puerto Rico. The NS&T Program consists of seven major components including the Benthic Surveillance Project, the Mussel Watch Project, and the Quality Assurance (QA) Project.

As part of the Benthic Surveillance Project, concentrations of organic and inorganic contaminants are determined in sediments and livers of bottom-dwelling fish taken in the same area. This Project quantifies environmental conditions at approximately 100 sites located around the nation. The analytes include polycyclic aromatic hydrocarbons (PAHs); polychlorinated biphenyl (PCB) congeners; DDT, its metabolites and other chlorinated pesticides; organotins; and major and

Benthic Surveillance Project are performed by the NOAA National Marine Fisheries Service's Northwest Fisheries Science Center, Seattle, WA, and Southeast Fisheries Science Center, Beaufort, NC. The same contaminants are determined in sediments, and mussels or oysters, as part of the Mussel Watch Project with the exception of mollusk tissues where PAHs are also quantified. The bivalves are collected usually on a yearly basis from approximately 200 sites in the United States, while sediments are collected at the same sites on a less-than-yearly basis. Sample collection and analysis for the Mussel Watch Project are currently performed by the Texas A&M University Geochemical and Environmental Research Group, College Station, TX, and the Battelle Laboratories at Duxbury, MA, and Sequim, WA.

2   QUALITY ASSURANCE PROJECT

The quality of the NS&T analytical data is overseen by the QA Project, which is designed to assure and document the quality of the data; to document sampling protocols and analytical procedures; and to reduce intralaboratory and interlaboratory variation. To document laboratory expertise, the QA Project requires all NS&T Program laboratories to participate in a continuing series of intercomparison exercises utilizing a variety of materials. The organic analytical intercomparison exercises are coordinated by the National Institute of Standards and Technology (NIST), and the inorganic exercises by National Research Council (NRC) of Canada.

The Environmental Protection Agency's Environmental Monitoring and Assessment Program (EMAP), is a complementary monitoring program to NOAA's NS&T Program. The EMAP is designed to evaluate the overall health of the Nation's ecological resources. The estuarine component of EMAP, EMAP-EC, is working closely with NS&T in monitoring estuarine ecosystems. To assure data compatibility, the EMAP-C laboratories participate in the intercomparison exercises. NOAA and EPA jointly provide financial support for the QA Project.

Approach

The NS&T QA Project is performance-based and no analytical methodology is currently specified. Laboratories can use any analytical procedure as long as the results of the intercomparison exercises are within certain specified limits of the certified or consensus values of reference or control materials. This allows the use of new or improved analytical methodology or instrumentation without compromising the quality of the data sets, and encourages the participating laboratories to use the most cost-effective methodology. Such a change was suggested by the NS&T laboratories and accepted by the Program office for the analysis of PAHs. The change was adopted after it was determined that the quality of the data generated using the new methods was comparable or superior to that generated using the old methods.

Methodology Documentation. All analytical methodology and sampling protocols used by NS&T Program monitoring projects are being documented.[1] The document is designed for use by both environmental science managers and laboratory chemists. The first part of the document provides the NS&T Program rationale, field collection methods for all sample types collected, QA Project criteria and goals, synopsized chemical analytical protocols, and gross pathology and histopathology methodology. Changes in the NS&T suite of analytes, method detection limits (MDLs), and MDL calculations through time will also be included. The second part of the document contains detailed descriptions of all the chemical analytical methods used by laboratories participating in the NS&T Program. These methods include those used since Program initiation to the present, for the determination of major and trace elements, and organic compounds in sediments and tissues.

The most common users of this kind of information are other organizations that have existing monitoring programs or are initiating monitoring programs. For those interested in interpreting the NS&T data or using these data to augment their own environmental characterization data, the information in the methods document supplies the kind of detail that will allow the researchers to determine whether or not other data sets are comparable to the NS&T data set. Through time, analytical methodologies will continue to improve as will the capability to differentiate contaminants in environmental samples. Documentation of NS&T laboratory analytical methods will allow scientists of the future of determine what modifications must be made to their data be able to use NS&T data for the determination of temporal trends.

Monitoring Site Documentation. Detailed site descriptions with latitude and longitudes for all sites sampled by both the Benthic Surveillance and Mussel Watch Projects have been complied.[2] The document contains a short history on the evolution of the sampling effort, site selection criteria, sample collection logistics, site location data, and other pertinent information. Also included are tables that provide annual sampling information by sample type, site, and year; and maps to assist in the location of sites.

Reference Materials. The analysis of reference materials, such as NIST Standard Reference Materials (SRMs) and NRC Certified Reference Materials, and of control materials generated for use by NS&T Program laboratories, as part of the sample stream, is required. Analytical data of all control materials and reference materials are reported to the NS&T Program office. These data are stored in the same format and at the same time as sample data, to form a data set that is part of the NS&T database.

In response to the needs of the NS&T and EMAP-EC, these Programs have partially funded the production of eight NIST SRMs and seven internal standard solutions. Two of these SRMs are composed of natural matrices that match those used in the NS&T Program. The calibration solutions SRMs contain two concentration levels of the three chemical classes of NS&T analytes. The latter are used to facilitate the preparation of multipoint calibration curves. The internal standard solutions were prepared at the request of the NS&T contract laboratories and are provided free of charge. The SRMs are available for purchase through NIST.

Analytical protocol requirements

Internal Standards. In NS&T Program trace organic analytical procedures, internal standards are added at the start of the analytical procedure and carried through the extraction process, cleanup, and instrumental analysis. Acceptable recovery rates are ±50%. It is the analyst's responsibility to monitor recovery rates and to determine acceptability based on variation of these rates.

Accuracy. The results of calibration checks performed at the beginning and end of each typical sample string must be within ±30% of the accuracy-based values for single organic analytes in certified standards, and ±20% for spiked blanks, in order to consider the instrument used to be within calibration. All samples must be quantified within the calibration range. Quantification based on extrapolation is not acceptable.

Detection Limits. MDLs are calculated and reported annually on a matrix and analyte basis. The method used for calculating MDLs is that defined by EPA and is described in detail in the United States Code of Federal Regulations[3]. If the EPA method is not used or is modified, the procedure used for MDL calculation is described in detail.

Precision. There is an inverse relationship between sensitivity and precision[4]. In general, the precision, as a function of concentration, appears to be independent of the nature of the analyte or the analytical technique. The interlaboratory coefficient of variation at the 10 ppb level is expected to be approximately 30%, and attainment of this level of precision will require the best possible effort on the part of the analyst. Thus the acceptable limits of precision for organic control materials for NS&T analyses are ±30% on average for all analytes, and ±35% for individual analytes. These limits are more rigorous for elemental determinations.

These limits apply to those materials where the concentrations of the compounds of interest are at least 10 times greater than the MDLs for those compounds. The application of these guidelines in determining the acceptability of the results of the analysis of a sample is a matter of professional judgement on the part of the analyst, especially in cases where the analyte levels are near the MDL.

QA samples. A minimum of 8% of the typical organic sample string must consist of blanks, reference or control materials, duplicates and spike matrix samples. The use of control materials does not entirely replace the use of duplicates and spiked matrix samples. In particular, the use of spiked matrix samples may demonstrate poor recovery of specific analytes. This percentage may be reduced to 5% if blanks are not considered. A minimum of 2% of the standard inorganic sample string consists of calibration materials and reference or control materials. The results of the routine analysis of reference and control materials, and of the intercomparison exercises, are stored electronically as part of the NS&T database.

Intercomparison Exercises

All the NS&T laboratories are required to participate in yearly intercomparison exercises. Participation by non-NS&T laboratories on a voluntary basis has increased with time. NS&T encourages such participation by the marine environmental community and will continue to add laboratories as funding permits. Laboratories participating in the exercises for the first time can request, if desired, a set of simple analyte solutions. Once these solutions are successfully analyzed or if the laboratory did not request them, the exercise materials are sent to each laboratory. The exercise materials are sent with complete handling instructions and a diskette with the data reporting format. Calibration solutions are also provided to eliminate errors caused by variations in standards. The type and matrix of the samples change yearly. Sample types have included simple gravimetrically-prepared analyte solutions, freeze-dried sediments, extracted freeze-dried tissues, and cryogenically homogenized tissues. If problems are encountered during any of the phases of the intercomparison exercises, the laboratories can contact NIST or NRC for assistance. Details of all of the intercomparison exercises have been compiled.[5,6] The results of the trace organic exercises are discussed briefly in Cantillo and Parris (elsewhere in this publication).

The results of the intercomparison exercises are not intended to be a reflection of the absolute capability of a laboratory. Given more time and money, the methodology used may yield even lower detection limits or better precision. Beginning in 1988, the intercomparison exercises were designed to improve the methodology used by the NS&T laboratories, by isolating sources of variability, such as sample preparation and extraction, not solely to assess it. The degree of complexity of the intercomparison exercise materials has increased since 1988 and recent exercises using natural materials are being used to assess method performance.

Quality Assurance Workshops. The results of the intercomparison exercises are discussed among NIST, NRC, and the participating laboratories during the yearly QA Workshop held in late fall or winter. During such meetings, a consensus is reached between NIST, NRC, NOAA, and the laboratories as to the type of materials that will be used for the following year's intercomparison exercise.

Laboratory performance. Overall, the performance of "core" laboratories, those that have participated since the beginning of the QA Project intercomparisons exercises, has improved with time. It is not possible, however, to document this statistically since different types of materials are used each year, and the difficulty of the analyses has increased with time as the level of expertise of the participating laboratories has improved. Thus, possible analytical errors may be due to matrix interference, low analyte levels, and other variables that change from year to year. It is possible, however, to compare the performance of a representative "core" laboratory with that of a laboratories participating in the intercomparison exercises for the first time.

Typical results of trace metal intercomparison exercises are shown in Figure 1. For the determination of As in mussel tissue and sediment, the performance of the "core" lab is comparable to that of the NRC, while that of the "new" lab needs more improvement than that of the representative "core" laboratory. The results of the 1991 intercomparison exercises for trace metals in sediments and tissues show that there is a positive correlation between performance and previous experience (S. Berman, NRC, personal communication, 1991).

The same observation can be made based on the trace organic intercomparison exercise results. An enriched bivalve tissue extract in methylene chloride was used as part of the 1990 organic intercomparison exercise. The mean absolute percent errors of the analyses performed by NS&T "core" laboratories were lower than those of the analyses of laboratories participating in the exercises for the first time. Typical results are shown in Figure 2. The values reported by NIST are shown for comparison. The results of the NS&T contract laboratories are within the ±30% limit of acceptability.

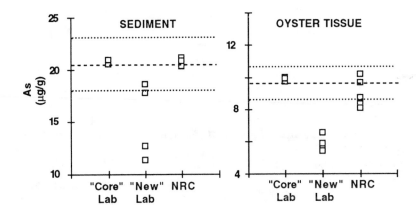

Figure 1 Results of the 1990 intercomparison exercise for As in sediment and oyster tissue. (Analysis of five replicates. The horizontal lines are the consensus mean value and ±20% of the mean.) (μg/g dry weight).

3   OTHER ACTIVITIES

Since its inception, the QA Program has participated in international quality assurance activities and coordinating groups. GESREM (Group of Experts on Standards and Reference Materials) is sponsored by the Intergovernmental Oceanographic Commission (IOC), the United Nations Environment Program (UNEP), and the International Atomic Energy Agency. GESREM has been mandated by its sponsoring agencies to prepare and distribute international standard and reference materials, and to disseminate information about such materials and their use. In response to this need, the NS&T QA Program compiled a catalog of reference materials for use in marine science. The third edition is currently available and lists close to 2000 reference materials from sixteen sources/producers worldwide.[7] The reference materials types listed in the catalog are ashes, gases, instrumental performance evaluation standards, oils, physical properties, rocks, sediments, sludges, soils, tissues, and waters. The catalog lists the producer/source, reference material description and preparation, analytes, certified and non-certified values, andcosts. Indices are included for

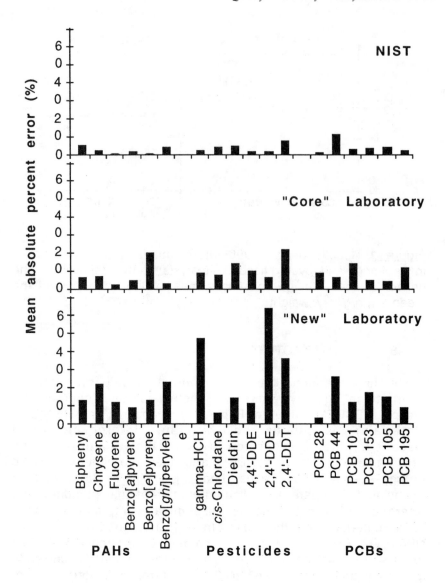

<u>Figure 2</u> 1990 Enriched bivalve tissue extract intercomparison exercise results mean absolute percent errors of PAH, pesticides and PCB analyses by NIST, an NS&T "core" laboratory and a non-NS&T laboratory participating in the intercomparison exercises for the first time.

locating elements, isotopes and organic compounds, as are cross references to Chemical Abstract Service registry numbers, alternate names and chemical structures of selected organic compounds. This catalog was published independently by both NOAA and IOC/UNEP. Copies are available from NOAA in printed or electronic form in Macintosh Word 4.0 format.

4   CONCLUSIONS

The quality assurance is an important part of environmental monitoring efforts. Performance based quality assurance not only maintains a status quo of interlaboratory comparability but allows for the introduction of new instrumentation and analytical techniques while increasing the accuracy and precision. In the NS&T Program, improved analytical methods were introduced for the identification PAHs, and laboratory precision was maintained at a high level or improved.

REFERENCES

1. G.G. Lauenstein, A.Y. Cantillo, and S.S. Dolvin, 1993, NOAA Tech. memo. (in preparation), NOAA/NOS/ORCA, Rockville, MD 20852, USA.
2. G.G. Lauenstein, M. Harmon and B.W. Gottholm, 1993, NOAA Tech. memo. (in preparation), NOAA/NOS/ORCA, Rockville, MD 20852, USA.
3. United States Code of Federal Regulations (1990) 40 CFR, Chapter 1, Part 136, Appendix B.
4. W. Horwitz, L.R. Kamps and K.W. Boyer, J. Assoc. Off. Anal. Chem., 1980,.63, 1344.
5. A.Y. Cantillo and R.M. Parris, 1993, NOAA Tech. Memo. (in preparation), NOAA/NOS/ORCA, Rockville, MD 20852, USA.
6. N. Valette-Silver, 1993, NOAA Tech. Memo. (in preparation), NOAA/NOS/ORCA, Rockville, MD 20852, USA.
7. A.Y. Cantillo, 1993, NOAA Tech. Memo. 68, 577 pp., NOAA/NOS/ORCA, Rockville, MD 20852, USA.

Evaluation of Trace Organic NOAA National Status and Trends Quality Assurance Project Performance

A.Y. Cantillo[1] and R.M. Parris[2]

[1]NATIONAL OCEANIC AND ATMOSPHERIC ADMINISTRATION, NOS/ORCA
N/ORCA 21, ROCKVILLE, MD 20852, USA
[2]NATIONAL INSTITUTE OF STANDARDS AND TECHNOLOGY, ORGANIC
ANALYTICAL RESEARCH DIVISION, GAITHERSBURG, MD 20899, USA

1   NOAA NATIONAL STATUS AND TRENDS PROGRAM

The National Oceanic and Atmospheric Administration (NOAA) has conducted the National Status and Trends (NS&T) Program since 1984 to determine the current status of, and any changes over time in the environmental health relative to toxic contaminants of the estuarine and coastal waters of the United States, including Alaska and Hawaii. The NS&T Program is a long-term monitoring effort, and since its inception in 1984, quality assurance has played a major role in the design of the program, the evaluation of potential analytical contractor laboratories, and the maintenance of data quality. The NS&T Program consists of seven major components: the Benthic Surveillance Project, the Mussel Watch Project, Bioeffects Surveys, Historical Trends Analyses, Regional Analyses, the Specimen Bank, and the Quality Assurance (QA) Project. The latter is described in detail in Cantillo and Lauenstein (elsewhere in this document).

The trace organic analytes determined as part of the NS&T Program include selected polycyclic aromatic hydrocarbons (PAHs); polychlorinated biphenyl (PCB) congeners; DDT and its metabolites; and other chlorinated pesticides (Table 1). The acceptable limits of accuracy for organic control materials for NS&T analyses are ±30% on average for all analytes, and ±35% for individual analytes. Sample collection and analyses for the Benthic Surveillance Project are performed by the NOAA/National Marine Fisheries Services/ Northwest Fisheries Science Center (NOAA/NMFS/NWFSC), Seattle, WA. The same contaminants are determined in sediments, and mussels or oysters as part of the

Table 1. Chemicals determined as part of the NOAA NS&T
Program

**Low molecular weight
PAHs** (2- and 3-rings)
1-Methylnaphthalene
1-Methylphenanthrene
1,6,7-Trimethylnaphthalene
2,6-Dimethylnaphthalene
2-Methylnaphthalene
Acenaphthene
Acenaphthylene
Anthracene
Biphenyl
Fluorene
Naphthalene
Phenanthrene

**High molecular weight
PAHs** (4-rings or larger)
Benzo[*a*]pyrene
Benzo[*b*]fluoranthene
Benzo[*e*]pyrene
Benzo[*ghi*]perylene
Benzo[*k*]fluoranthene
Benz[*a*]anthracene
Chrysene
Dibenz[*a*,*h*]anthracene
Fluoranthene
Indeno[*1,2,3-cd*]pyrene
Perylene
Pyrene

DDT and metabolites
2,4'-DDD   4,4'-DDD
2,4'-DDE   4,4'-DDE
2,4'-DDT   4,4'-DDT

Chlorinated pesticides
Aldrin
cis-Chlordane
Dieldrin
gamma-HCH (Lindane)
Heptachlor
Heptachlor epoxide
Hexachlorobenzene
Mirex
trans-Nonachlor

PCB congeners

PCB 8	PCB 18	PCB 28
PCB 44	PCB 52	PCB 66
PCB 101	PCB 105	PCB 118
PCB 128	PCB 138	PCB 153
PCB 170	PCB 180	PCB 187
PCB 195	PCB 206	PCB 209

Mussel Watch Project. Sample collection and analysis for the Mussel Watch Project are currently performed by the Texas A&M University Geochemical and Environmental Research Group, College Station, TX, and the Battelle Laboratories at Duxbury, MA. The quality of the NS&T analytical data is overseen by the QA Project, which is designed to assess and document the quality of the data; to document sampling protocols and analytical procedures; and to reduce intralaboratory and interlaboratory variation.

2   INTERCOMPARISON EXERCISES

To document laboratory expertise, the QA Project requires all NS&T Program laboratories to participate in a continuing series of intercomparison exercises utilizing a variety of materials. Currently, the organic analytical intercomparison exercises are coordinated by the National Institute of Standards and Technology (NIST), and the inorganic exercises by National Research Council (NRC) of Canada. The 1986 organic exercise, however, was coordinated by the NOAA/NMFS/NWFSC laboratory in Seattle, WA. The results of the intercomparison exercises are fully documented.[1] An increasing number of non-NS&T laboratories are participating in the intercomparison exercises on a voluntary basis.

<u>Exercise Materials.</u> The exercise materials are distributed in the spring or early summer, and the final results are collated and evaluated in late fall just before the annual QA Workshop. Laboratories participating in the exercises for the first time can request, if desired, a set of simple analyte solutions. Once these solutions are successfully analyzed or if the laboratory did not request them, the exercise materials are sent to each laboratory with complete handling instructions and data reporting format. The type and matrix of the exercise materials change yearly in response to specific needs identified by the participating laboratories (Table 2). Exercise materials have included simple gravimetrically-prepared analyte solutions, wet and dry sediments, extracts of tissues, and cryogenically-homogenized frozen tissues.

<u>1986.</u> The materials used for the 1986 exercise were a contaminated sediment from the Duwamish Estuary, Seattle, WA, and a mussel tissue homogenate from specimens artificially exposed to contaminated water. The Duwamish sediment material had been characterized extensively prior to use in the NS&T program but that information was not given to the analysts. The results for the summed low molecular weight (2- and 3-ring) PAH concentrations showed good precision and were within ±30% of the grand mean of the six individual laboratory means. The results for the summed high molecular weight (4- and 5-ring) PAH concentrations, however, showed large discrepancies in precision and accuracy. Results of limited determinations of PAHs in these materials by NIST using both gas chromatography and liquid chromatography were

<u>Table 2</u> Materials used for the NS&T trace organic intercomparison exercises

Year   Materials

<u>1986</u> (organized by NAF)
DWAMISH   Duwamish Estuary sediment, Seattle, WA
Ti86   Mussel tissue from Puget Sound, WA
QA   Calibration: Standards in Hexane

<u>1987</u> (organized by NIST)
Sed87   Sediment from Baltimore Harbor, MD
Ti87   Mussel tissue from Narragansett Bay, RI
QA   Control Materials: Baltimore Harbor sediment and mussel tissue from Narragansett Bay
                 Calibration: NIST SRMs 1491, 1492, and 1493, and other solutions used as internal and surrogate standards

<u>1988</u> (organized by NIST)
Var-PAH   PAHs in hexane and toluene
Var-PCB   PCBs in 2,2,4-trimethylpentane
Var-PES   Pesticides in hexane
Ti88   Mussel tissue from Boston Harbor, MA
QA   Control Materials: Baltimore Harbor sediment, mussel tissue from Narragansett Bay, and mussel tissue from Boston Harbor
                 Calibration: NIST SRMs 1491, 1492, and 1493

<u>1989</u> (organized by NIST)
Var2-PAH   PAHs in hexane and toluene
Var2-PCB   PCBs in 2,2,4-trimethylpentane
Var2-PES   Pesticides in hexane
QA89T1   Oyster tissue
ICES PCBs   First exercise
QA   Control Materials: Baltimore Harbor sediment and mussel tissue from Boston Harbor; SRM 1941
                 Calibration: NIST SRMs 1491, 1492, 1493, 2260, 2261 and SRM 2262

<u>1990</u> (organized by NIST)
Var3-Mix   PAH, PCB and pesticides mixture in hexane
QA90E1   Enriched bivalve tissue extract
ICES PCBs   Second exercise
QA   Control Materials: Baltimore Harbor sediment, mussel tissue from Boston Harbor, and tissue control material III; SRMs 1941, 1974
                 Calibration: NIST SRMs 1491, 1492, 1493, 2260, 2261, and 2262

consistent with each other and the variations observed were typical of those observed when using two different procedures. The NIST results agreed reasonably well with the exercise results from the other participating laboratories. Large discrepancies between the NIST values and those of the NS&T laboratories were observed for fluoranthene and pyrene. For each laboratory, the standard deviations of the PAH analyses replicates were between 2 and 23% except for the results of one analyte by one laboratory. Little clustering was found according to analytical method used. This shows that differences in analytical methodology had minimal effect on the analytical results. The summed PCB data for each NS&T laboratory was within ±30% limits of the grand mean. The standard deviations of the individual congener analyses replicates were below 30% except for analytes present in very low concentrations. The pesticide concentrations reported were very low, and many were below the method limit of detection (MDL). With little data, no consensus means could be calculated and no further analysis of the pesticide data was performed.

The results of the 1986 mussel material analyses of the low and high molecular weight PAHs were within ±30% of the grand mean of the individual laboratory means. As in the case of Duwamish sediment, the grand means and the ±30% limits were used as guides as there was no true or consensus values. The 1986 mussel tissue data submitted by some of the laboratories was less precise than that of the Duwamish sediment, with standard deviations of replicate analyses ranging overall up to 30% with some higher exceptions. This difference in precision was due to effects of the more complex natural tissue matrix and had been observed previously. The summed PCB data for five of the NS&T laboratories were within ±30% limits of the grand mean, and only one laboratory reported data below the lower limit. Several laboratories again failed to report values for some analytes, or reported values below the MDLs.

1987. The materials used were fresh frozen mussel tissue homogenate and a sediment, both prepared by NIST. The concentrations of PAHs in the sediment material were low and although some of the data reported consisted of values below the MDLs, most PAHs on the NS&T analyte list were quantified. As with mussel tissue analyses in 1986, there were various instances of observed peak interference for the PCB congener analyses, and possible misidentification of peaks in the gas chromatograms. Values were not reported for many analytes or were below the MDLs. The pesticides analysis results of the sediment material were very low and close to the MDLs and results for many analytes were not reported.

The concentrations of PAHs in the tissue material were very low and most of the data reported consisted of values below the MDLs. No conclusions could be made from this portion of the exercise. There were again various instances of observed peak interferences and possible misidentification of peaks in the gas chromatograms of the PCB congener analysis, and values for many analytes were not reported or were below the MDLs. The pesticides analysis presented similar problems to those encountered for the PCB analysis. Peak interferences were noted between the pesticides and PCB congeners by several laboratories.

1988. Due to the problems encountered during the 1987 exercise, a series of solutions were used for the 1988 exercise instead of the more complex natural matrix materials. These solutions were designed to help isolate the sources of variability and bias in the analyses, such as the instrumental analysis step. Three solutions were prepared gravimetrically to mimic the relative analyte concentration range found in real samples. The solutions were PAHs in hexane and toluene, PCBs in 2,2,4-trimethylpentane, and pesticides in hexane. Each solution contained all the analytes in each chemical class listed in Table 1, and the list of analytes was provided to the laboratories. The laboratories were asked to analyze the contents of one of the ampoules in triplicate (i. e., three GC injections), and the other two ampoules were analyzed once. The results of the 1988 exercise demonstrated that the NS&T laboratories could identify the compounds in the exercise materials chromatograms and could quantitate them within the limits specified by the NS&T Program. The accuracy and precision, however, could be improved.

1989. The materials used were gravimetric solutions (PAHs in hexane and toluene, PCBs in 2,2,4-trimethylpentane, and pesticides in hexane); and a frozen oyster tissue homogenate of specimens collected in the Gulf of Mexico. The list of analytes in the gravimetric solutions was provided to the laboratories. The laboratories were asked to analyze one of the extracts in triplicate, and the other two once. All the PAH, PCB and pesticides analyses results were within the ±35% limits with the exception of the 2,4'-DDD results of one laboratory. The results of the three solutions showed good precision and accuracy. The oyster tissue deteriorated to a certain degree before being received at NIST. The material presented severe analytical problems and the results of the exercise could not be used as a measure of laboratory performance.

<u>1990</u>. The exercise materials were a gravimetrically-prepared solution of six aromatic hydrocarbons, six chlorinated pesticides and six PCB congeners in 2,2,4-trimethylpentane; and an enriched bivalve tissue extract in methylene chloride. After clean-up, the laboratories were asked to analyze one of the extracts in triplicate, and the other two extracts once. The tissue extract was enriched for the same 18 compounds present in the 2,2,4-trimethylpentane solution. The 2,2,4-trimethylpentane solution was used to test the ability of the laboratories to separate, identify and quantitate individual compounds. The NS&T laboratories correctly identified the 18 compounds and the RSDs were well below 10% for most cases.

The results of the 1990 intercomparison exercises are summarized graphically in Figures 1 and 2, and the order of analytes is shown in Table 3. For each laboratory and analyte, the ratio of the gravimetric or consensus value to the mean reported concentration from the analyses of three samples was calculated. In order to keep relative differences on a common scale, ratios were forced to be 1.0 or greater by always dividing by the smaller number. Thus, the limit of the established acceptability limit of ±35% would result in a ratio of 1.35. All the results are close to or below 1.35, with the exception of benzo[*a*]pyrene and benzo[*e*]pyrene in both exercise materials, and of PCB 195 in the enriched tissue extract results for one laboratory.

3   RESULTS AND CONCLUSIONS

The QA Project is an essential part of the NS&T Program as it assesses data quality, documents the performance of the laboratories and promotes a continued high level of performance. The performance of the NS&T laboratories doing trace organic analysis has improved since the beginning of the Program. Since it is not possible to document the improvement in performance of laboratories participating in the

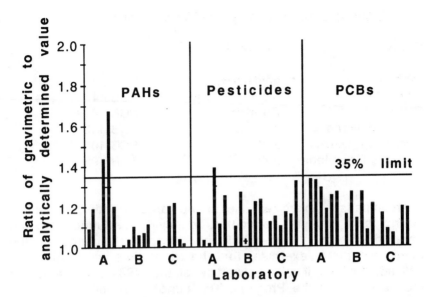

Figure 1 1990 Gravimetric solution mixture intercomparison exercise ratios of gravimetric to analytically determined mean value of three samples by laboratory. (Ratios forced to be >1. Order of analytes same as in Table 3. + - not reported.)

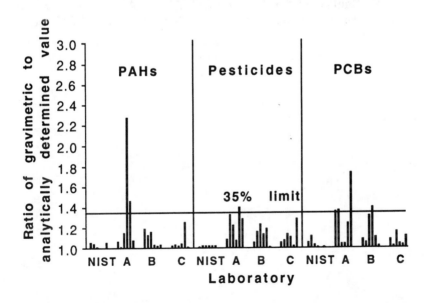

Figure 2 1990 Enriched bivalve tissue extract intercomparison exercise ratios of consensus value to analytically determined mean value of three samples laboratory. (Ratios forced to be >1. Order of analytes same as in Table 3.)

<u>Table 3</u> Order of analytes in Figures 1 and 2

PAHs	Pesticides	PCBs
Biphenyl	gamma-HCH	PCB 28
Fluorene	*cis*-Chlordane	PCB 44
Chrysene	Dieldrin	PCB 101
Benzo[*e*]pyrene	4,4'-DDE	PCB 153
Benzo[*a*]pyrene	2,4'-DDE	PCB 105
Benzo[*ghi*]perylene	2,4'-DDT	PCB 195

intercomparison exercises due to the large variation in sample matrix and analyte concentration level, evaluation of the performance of a non-NS&T laboratory participating in the intercomparison exercise for the first time may give an indication of the level of expertise of the NS&T laboratories at the beginning of the Program. The performance of a non-NS&T laboratory just joining the NS&T QA Program is typically not as good as that of NIST or of an NS&T laboratory.

4   REFERENCE

1.   A.Y. Cantillo and R.M. Parris, 1993, NOAA Tech. Memo. (in preparation), NOAA/NOS/ORCA, Rockville, MD 20852, USA.

Operating under US FDA and US EPA Good Laboratory Practices

James A. Ault

ABC LABORATORIES, INC., PO BOX 1097, COLUMBIA, MO 65205, USA

1 INTRODUCTION

As a contract laboratory, ABC Laboratories has operated under the Good Laboratory Practices (GLP's) since 1978, when the US FDA promulgated the first such regulation. The purpose of this conference is to compare and harmonize GLP regulations across international boundaries. As such, the presentation will not concern itself with each segment of the regulations, but rather with those areas where compliance has been problematic.

Study Director

Due to the definition of study director in the various regulations, the position is controversial. Since the management at each facility must delineate the qualifications and oversee the responsibilities, different criteria may exist even within one facility.

Qualifications, for example, may be different for the study director between departments. The chemistries required to accomplish basic water chemistry and verify concentrations of a test substance in water are far less stringent than those required to determine metabolic pathways in animal systems or degradates in environmental samples. At our laboratory, we follow the US Office of Personnel and Management guidelines for entry-level qualifications, and insist on documented training and experience in like positions before bestowing the title. It generally takes from three to five years experience on the job to qualify for a position.

Compounding the difficulty of qualifications, the recent US EPA redefinition of study director requires responsibility for all phases of very broad studies (field studies in particular). We can no longer require study directors to be expert at all phases of the study because of the diversity of disciplines within these studies. Reliance on other professionals at other facilities has become the watchword, and the study directors are forced to travel with the quality assurance auditors to maintain that confidence. Reporting relationship within a study can be tortuous and convoluted in many ways, requiring the utmost of management skills to coordinate properly. A typical magnitude of residues study may involve the sponsor, four to eight field sites and three to six analytical laboratories to accomplish. The study director may reside at the sponsor, a field site or one of the laboratories.

As with all contract facilities, we must concern ourselves with the workload of each study director at the laboratory. The day-to-day maintenance of a study can be delegated, but weekly review of results and oversite cannot. Report writing, monitoring results and comparing to the protocol or method requires time, and an overload of studies will not produce the quality or timeliness promised.

Test Compounds

All GLP's require characterization of the test compound before use in a study. As a contract facility, we receive many test compounds from many sources. Chain of custody requirements have led us to develop a computer database for test compound accountability which includes all elements of audit trail while a chemical is at our facility. Our Material Compliance program enhances test compound supervision, traceability, Hazard Flagging System, and material safety data sheet administration. Tracing a compound to the source records and identifying the location of characterization data (offsite or onsite) is an easy matter with this system. In addition, identifying which test compound "empty containers" can be discarded is also enhanced.

Laboratory and Field Notebooks

The most difficult part of compliance with the GLP's is probably the collection and archiving of notebooks which contain original raw data. While we do not use notebooks for observations or operation of our studies, a set of records which we call "facility" raw data are recorded in instrument logs, culture records and water system logs. These records are of a continuous nature, and originals cannot be sent with the raw data package of any one study, because they will be used for a number of studies over time. If a "facility" record needs to be included in a report, we will make an exact, certified copy with a date stamp for inclusion in the report. The originals are maintained in the archive facility. Verifying collection of these records is probably the most difficult responsibility required of the QA Unit, and we have included a routine monitoring of these records in all audit procedures to assist us in collecting those logs that are complete, and to ensure that they are archived properly.

Computer Validation

With the issuance of the "guidelines" for computer validation, serious efforts have been expended to define and refine our program to comply with the Good Automated Laboratory Practices (GALP's). ABC Laboratories formed a GALP Committee comprised of Computer System, User and QA Unit personnel to address our needs and formulate strategies for compliance. As with any facility, we have treated personal computers (PC's) and Systems as separate pieces of the whole. PC software programs cannot be validated; thus our approach is to verify 100% of all data and calculations generated.

In the Systems area, any programs that generate or collect raw data have been identified and either validated or are in the process of being validated. Much of our difficulty has been with the audit trail not being documented well enough. Our vendors are generally responsive to requests for assistance, but the ultimate burden falls on us to validate. In many cases, a test set of data can be used repeatedly (once it is verified) to prove validity, but the set must be designed to use every program and hardware piece needing validation (sometimes not an easy task).

2 SUMMARY

Since this conference is to compare and harmonize regulations across international boundaries, the topics discussed in this presentation have been concerned with problems encountered within the set of GLP regulations of the US FDA and EPA. Solutions that have been implemented are offered, in the hope that a "harmonized" set of GLP's may address those issues.

Management and Motivation in Quality Assurance: The Human Element in Proficiency

Trean Korbelak Blumenthal

LIBRA LABORATORIES, INC., 16 PEARL STREET, METUCHEN, NJ 08840, USA

"Quality is never an accident. It is always the result of intelligent effort.
There must be the will to produce a superior thing." This is a quotation from
John Ruskin, a 19th-century English writer and social reformer. He points out
that motivation is essential for quality assurance.   The "will to produce a
superior thing" is related to the value system of both individuals and
organizations. Achieving a good fit between the values of the individual and the
needs of the organization --in this case, the analytical laboratory-- is a crucial
factor in optimizing motivation and making work life more successful and more
enjoyable for both scientist and manager.

Quality assurance is an organizational investment strategy, and this is no less
true in the human dimension. Management systems which select for, develop,
support, and reward the drive for excellence, achievement, recognition, and self-
expression are more likely to achieve quality assurance goals.   Good
management practices and good lab practices go hand in hand in optimizing
laboratory quality and performance.   Lab managers sometimes tend to
concentrate almost exclusively on technical issues, but there are advantages to
be gained in actively and creatively managing the people issues of the lab. Some
points to consider in management for motivation to quality are:

1. Hire people who love to do the work you need them to do.

It takes a lot of internal motivation to complete a degree in chemistry.
Make sure the person you hire is enthusiastic about doing analytical chemistry,
and prepared for the discipline it requires. People do best at work they enjoy for
its own sake, and not just as a way to make a living. Consider also the impact of
the new person on the existing lab group, and vice versa, to assure compatibility.
A manager with good interview skills can learn about a candidate's natural
motivations and help make a good match to the organization's needs.

2. Train people well; provide them with good tools; support their continuing development; and give them a sense of purpose in their work.

Proper training builds skill and confidence; and continuing development
opportunities are both a reward for successful accomplishment and a natural
response to the human desire for personal growth. This training should go
beyond analytical measurement skills, lab safety, sample handling, statistical
analysis, lab Q.A., and documentation and reports. Analysts should know the
mission of their lab, who its "customers" are, and the nature and value of the
decisions made as a result of the data the lab generates. An analytical result has

much more meaning if it is deeply understood that someone's life or health are at stake, or that a decision will be made about whether millions of dollars worth of goods will be used or wasted, based on that measurement result and its reliability.

How interesting it would be to see a lab's accomplishments reported in terms of economic impact, value of goods safeguarded, etc., rather than just a record of numbers and types of samples analyzed. Knowing the organizational and societal implications of one's work gives it context and value, and increases the prestige of the lab and all who work in it.

3. Make sure people clearly understand their work tasks and the desired outcome.

Avoiding ambiguity about job responsibilities and standards of performance prevents hesitation and uncertainty in the conduct of the work. Assignments should be given at a level which is consistent with the analyst's skills, but also offers challenges to strengthen and expand skills, in order to build confidence and foster a sense of accomplishment. Use the lab's Q.A. program as a source of information on how people are doing and whether they have achieved mastery.

4. Provide regular feedback and recognition of accomplishments.

Feedback should not await a formal performance review process, which all too often is an annual appraisal feared and dreaded by all participants. People need to know how they're doing as they go along, and should never be surprised at their annual review. An ideal feedback mechanism is one which is discoverable by the analyst himself/herself, which regularly gives reinforcement of good practices by demonstrating competency. Here again, the lab's technical Q.A. program can provide an objective and self-evident way for people to get feedback on their work.

A kind of "peer review" process, in which journeyman level analysts help each other and junior analysts to review and correct work, can be helpful and comfortable as a feedback device, and developmental for both junior and senior level personnel.

Supervisory and management feedback must be fair and consistent with a recognized and appropriate performance standard. It should be both corrective and supportive, not just fault-finding. "Catch them doing something right", for positive reinforcement of good practices. This casts the supervisor/manager as an ally, not a critic, and builds confidence and trust in the relationship. A little sincere encouragement goes a long way. Value your people and make sure they know it.

5. Develop individual and team diagnostic and problem resolution skills.

One area of skill development which is often overlooked is that of diagnosis and resolution of analytical lab and methods problems. Since analytical chemists are by nature orderly and systematic in their work, they are very well suited to "detective work". Involving the analyst in individual and collaborative problem identification, analysis, and resolution gives unique insights into critical control points in analytical procedures and lab operations, resulting in better quality work. It also builds teamwork and a cooperative spirit in the lab, while taking away from management some of the burden of solving every problem.

Such work lends variety to the work day and builds confidence and self-esteem. Widespread capabilities of this kind also help to deter any tendencies to fraud and misconduct in the production of measurement data, since these can now be more readily detected.

Group brainstorming and other creative problem-solving techniques can also be used to good effect to draw out the staff on issues of concern and on better ways to do things, using the synergy of the group.

6. Rotate work assignments for variety and growth.

The opportunity to vary one's work can be an important motivator. Because of the rigorous control required to do analytical chemistry, it is important to assure some variety, freedom, and creative outlets for the analyst. The day-to-day measurement process requires strict adherence to prescribed procedures. People are creative and imaginative and may try to avoid boredom and "improve" things. In order to prevent these "creative improvements" --i.e., deviations from specified methods-- from occurring in the middle of controlled analyses, it is useful to give analysts some time to work in areas where they can constructively use creativity and imagination. For example, analytical methods development assignments, training of junior staff, design of more effective/efficient report forms, validation study work, and Q.A. systems design and implementation, all offer opportunities to make useful contributions to the lab, while getting away for a little while from the constraints of what is often repetitive work.

Involvement in creating new procedures and training others forces the analyst to look at his or her work in a new way and organize it so others can do it as well. This improves the analyst's own bench work and also provides growth toward supervisory/management roles.

7. Be creative in the rewards system.

While financial compensation is an important basic consideration in rewarding work well done, scientists have other human and intellectual needs, and these can form the basis for motivational management. Prestige among colleagues, chances to learn new things and develop to full potential, chances to share what one knows, visibility within the larger organization beyond the lab, opportunities for recognition and advancement, and many other hopes, wishes, and needs, can form the basis for stimulating and meaningful reward systems, such as:

a. Attending appropriate technical and professional meetings.

b. Giving presentations and publishing. These may be technical papers at professional society meetings, or they may be talks and/or reports given within an organization, e.g., where proprietary issues are involved. Effective seminars and presentations enhance the image of both the individual and the lab unit.

c. Continuing education support, whether for formal academic study for an advanced degree, or short courses designed to expand knowledge of a specific subject.

d. Involvement in training others, in-house consulting work, technology transfer, and project work and program design and implementation.

All of these rewards offer challenge, responsibility, recognition, advancement, and personal growth which are part of the motivation of the professional scientist and analytical chemist.

Conclusion

All of these motivational methods require a management investment. Today's harried lab manager may despair at finding time and energy for making such investments. The rewards, however, are tremendous -- for the manager, the scientific staff, the operating unit of the laboratory, and the larger organization in which the lab resides. A manager's legacy is the success of his or her staff to achieve great goals and to create effective systems which outlive them all. Recognition of the human element in proficiency can help with both day-to-day quality assurance and these longer-range goals and aspirations.

A child was once asked what was the best reason for Handel to work so hard composing music, and the answer was: "The sound of everybody shouting 'bravo!'". What greater motivation than this?

Quality Assurance in the Hazardous Waste Industry: A Multiple Laboratory Quality Control Program

Eugene J. Klesta
CHEMICAL WASTE MANAGEMENT INC., ALSIP, IL 60658, USA

INTRODUCTION

With the enactment of the Resource Conservation and Recovery Act (RCRA), the United States hazardous waste industry was required to track waste materials from "cradle to grave." Generators of waste are required to characterize the waste by specifying the chemical constituents of the waste. Companies which treat, store, and dispose of waste, commonly referred to as TSDFs, are required to perform these actions properly. Analytical information is necessary to make the correct management decisions. The analysis of hazardous waste is very difficult because of heterogeneity of the material and the high variability of the waste stream itself. Generators do not have specifications for the waste streams which they generate. Their primary concern is to remove the waste from their facility and to have it managed in an environmentally conscious manner.

COMPANY BACKGROUND

Chemical Waste Management Inc. (CWM) is the world's largest hazardous waste company and has developed a network of analytical laboratories distributed across the United States. The company gathers information about the waste to determine whether or not it should be treated, stored, or disposed of properly. Health and safety concerns and operating permit conditions are addressed by knowing what chemical constituents are present in the waste. To ensure uniform decision-making processes and management techniques, CWM developed a quality assurance and quality control program which took into consideration the fact that multiple laboratories were going to be included in the program. CWM used the Good Laboratory Practices (GLPs) that are found in the Fungicide, Insecticide, Fumigants, and Rodenticide Act (FIFRA) and the Toxic Substances Control Act (TSCA) for the basis of its quality assurance program. GLPs require specific quality assurance procedures along with the generation of defensible documentation to ensure the safety of the general population. CWM used these two documents as guidance for developing its rigorous QA/QC program.

QUALITY CONTROL PROGRAM

The quality control program was designed to be implemented in a variety of laboratories across the country, from large laboratories with excess of 100 analysts to small laboratories that may be staffed by two or three analysts. The program is parameter-driven rather than batch or sample driven. This means that each analytical parameter that is required for a waste stream causes certain quality control procedures to be performed. The general flow of the quality control requirements for a single parameter includes the following steps:

1. Instrument Performance Check

    An instrument performance check must be performed for each piece of analytical equipment on a daily basis. The instrument performance check is a measurement which can be determined rapidly and indicates whether or not the instrument is operating within the acceptance criteria for that particular day. Statistical analysis is done on the measurements; criteria are established; and each analyst uses the daily measurement to determine if the analytical process should continue or corrective action is needed.

2. Blank Analysis

    Analysis of a method blank is performed with each batch of samples. This procedure monitors the absence of cross contamination from one sample to the next or occurrences of external contamination. Because many of the samples which CWM handles are highly concentrated in certain chemical constituents, it is necessary to ensure that the results for a particular sample represent that sample and not the one before it in the analytical sequence. Criteria for the analysis of blanks are established. No analysis is approved or reported if the blank analysis exceeds the reporting limit for that particular parameter.

3. Quality Control Check Sample

    Each laboratory is required to analyze a quality control check sample for each parameter on a daily basis. These results are compiled on a computer system, quality control charts are developed, and the criteria are used to determine whether or not the results are acceptable. Warning limits are at $\pm$ 2 standard deviations and control limits are established at $\pm$ 3 standard deviations of the mean. The quality control check sample is carried through the entire analytical method. Check samples are either purchased by the laboratory or are produced within the laboratory by means of weights and volumes.

4.   Duplicates

Because of the wide variety of waste streams submitted to a CWM laboratory, a duplicate analysis is performed on every tenth sample for each parameter. A duplicate is defined as two test portions taken from the sample, which was submitted to the laboratory, and carried through the entire preparation, extraction, and determination steps of the method. The criterion for acceptability of duplicate analysis has been established at less than 20% relative error.

5.   Fortification

Fortification analysis is also performed on every tenth sample. A third test portion of the waste sample is taken and a known amount of analyte is added to the test portion. Using the theoretical amount and the results for the fortification analysis the percent recovery is calculated. Criteria for percent recovery are 80% to 120%. Both duplicates and fortifications are tracked using control charts.

These five procedures and the associated criteria would constitute quite a rigorous quality control program by themselves. The CWM laboratory managers use the results from these procedures to monitor precision and accuracy within the laboratory and to ensure that the analytical systems are in control.

BLIND DUPLICATES

The quality control policy requires that each laboratory performs blind duplicates on a weekly basis. This procedure produces additional information that the laboratory manager uses to ensure that all instruments are performing in a proper manner and that all analysts are generating comparable results. The laboratory manager assigns a sample of a particular waste stream to an analyst which was previously analyzed in the laboratory. It is then possible to estimate the intra-laboratory precision based upon the results from the first analysis compared to the results of the second analysis. Although the second analyst knows that the material being analyzed has been analyzed previously and is labeled as a "blind duplicate", information as to the source of the sample or the concentration of the constituents in that particular material are unknown.

Many programs incorporate "double blind" samples. The analysts do not know the concentration of the analytes and are unaware that the sample is, in fact, a standard or a previously analyzed sample. To set up a "double blind" program within a large multiple laboratory organization would be extremely complicated. A significant amount of information would have to be generated to keep the analysts totally unaware of the source of the samples. The use of blind duplicates, as described above, is adequate because of the frequency required by the CWM QC Policy.

CWM incorporated two additional practices to ensure that the analytical data generated in any CWM laboratory would result in the same disposal decisions and the same waste management techniques for the material. These two components of the CWM program are probably unique in the hazardous waste industry.

ROUND ROBIN PROGRAM

A proficiency testing scheme was developed in 1984 to evaluate the performance of CWM laboratories analyzing "real world" samples. There were many water matrix proficiency programs in the United States at the time, but none of these were applicable to the hazardous waste industry because a significant amount of waste is either sludge or solid. CWM developed its program in a niche of its own. The samples distributed are restricted to hazardous waste materials (solids, liquids, and sludges). The analyses requested include hazardous waste constituents such as cyanide, polychlorinated biphenyls (PCBs), metals, and a variety of other parameters that are used to characterize the waste and to handle it properly. The quality assurance unit, located in Alsip, Illinois, prepares these materials on a quarterly basis. The concentrations of the analytes are determined by weights and volumes and are unknown to the participating laboratories.

The samples are shipped to the laboratories with instructions and a floppy disk which contains a report form specific to the particular quarter. The use of these disks for reporting data makes the compilation of data and the statistical report generation very efficient.

The analytical methods used are typically taken from the United States Environmental Protection Agency (USEPA) solid waste methods manual referred to as "SW-846". The laboratories are given approximately five weeks to analyze the samples and report the information back to the quality assurance unit. In addition to the CWM laboratories, all outside laboratories, which are approved by the quality assurance unit, are also included in this round robin program. The total number of laboratories participating in the program is now ninety (90), but has been as high as one hundred and twenty (120).

The analytical results are compiled, normal probability plots are generated and reviewed, the Grubbs test is used to determine and remove statistical outliers, and the mean and standard deviation are determined from the remaining data. The mean and standard deviation are used to calculate the Z-score for each analytical result. All laboratories with a Z-score within $\pm$ 2 are considered to be acceptable. Those outside of two units are considered to be unacceptable and corrective action must be taken. Once the report is sent back to the laboratory, it is their responsibility to investigate the causes for any outliers or any results which fall outside the acceptance criteria. Possible causes may include the analyst, the instrument, or the lack of adherence to the method.

One of the major benefits of this program is that it gives CWM the ability to assess the accuracy of the analytical results being done at its 38 laboratories. This program gives the company the ability to approve outside laboratories based on performance. The round robin results also demonstrate analytical accuracy when the calculated mean approximates the theoretical concentration.

REFERENCE LABORATORY EVALUATION PROGRAM

The second major quality control procedure, used for a multiple laboratory operation, is the Reference Laboratory Evaluation program, which we commonly refer to as the parallel analysis program. This feature of the CWM quality assurance system is quite different than most other quality control programs. Each month, all the laboratories within the CWM system are required to submit one sample to the reference laboratory located in Riverdale, Illinois. The laboratory performing the original analysis submits the sample to the reference lab without the analytical data. The reference lab performs its analysis without knowledge of the concentrations of the analytes. The analytical data from the original laboratory is submitted to the quality assurance unit. The quality assurance unit then receives the corresponding analytical results from the reference laboratory and performs a review and comparison of both sets of analytical results.

The criteria used for the comparison of data are: (1) the analytical results from the two laboratories agree within 20% relative error and (2) the same disposal decision would be made at either location. Because of regulatory limits or permit conditions, it is possible to have analytical results within 20% relative error and not be able to manage the waste. This situation occurs when the results are on opposite sides of one of these legally imposed limits.

One of the additional features that makes this program valuable to the company comes from the fact that the analytical data are being compiled in a computerized relational database. The quality assurance unit has the ability to query the database and investigate any trends, laboratory bias, method bias, or problems with the comparability of the analytical data. Each of the samples that is submitted as a reference laboratory sample has all of the analytical data stored in the database. Quality assurance staff have the ability to compare results for a single parameter for many laboratories or the results for many parameters in a single laboratory through time.

The graph below demonstrates just one example of the usefulness of the parallel analysis program. It demonstrates how this database can be used to convert analytical data into information. By examining the results in order, we can make the following observations about each laboratory:

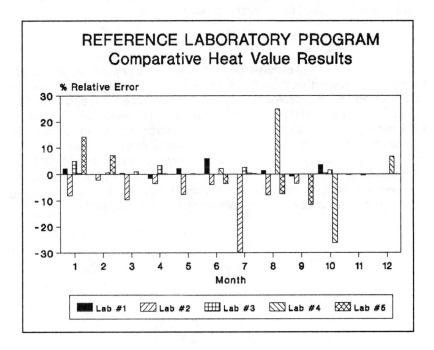

1.   Laboratory #1 demonstrates the expected results. The occurrences of positive and negative percent relative errors appears to be random and all of the results are below 20%.

2.   The data for laboratory #2 shows a significant negative bias in comparison to the reference laboratory. One of the results (seventh month) exceeded the acceptance criterion.

3.   The results from laboratory #3 show a slight positive bias, but all of the results are within the allowable range.

4.   During the eighth and tenth months, laboratory #4 exceeded the acceptance criteria and in opposite directions. Immediate corrective action was taken to locate the source of error.

5.   The results for laboratory #5 present the most interesting situation. A trend from a positive bias to a negative bias is obvious. Although all of the results are within the acceptance range, investigation would be conducted to find the source of this "unexpected" condition.

The parallel analysis program gives CWM a unique opportunity to examine results from a variety of laboratories across the country and to be sure that the same environmentally safe disposal decisions and treatment scenarios will be used for a particular waste stream.

CONCLUSIONS

Implementing all of these required quality control procedures at CWM laboratories means that the laboratory personnel must be properly trained to understand the rationale for the quality control procedures and to be able to interpret the quality control data for its maximum utilization. Determination of systematic or random error occurring within the laboratories can be found and corrected using these procedures. This internal quality assurance scheme is not only necessary, considering the high variability of the waste streams in the hazardous waste industry, but also results in a technically sophisticated mechanism for assessing precision and accuracy in a multiple laboratory organization.

Missing Aspects in Quality Control

Robert W. Dabeka and Stephen Hayward

FOOD RESEARCH DIVISION, BUREAU OF CHEMICAL SAFETY, FOOD
DIRECTORATE, HEALTH PROTECTION BRANCH, HEALTH AND WELFARE
CANADA, OTTAWA, ONTARIO, CANADA K1A 0L2

1 INTRODUCTION

Two groups of evidence demonstrate that analytical accuracy becomes poorer as the levels of analyte requiring determination approach ng/g and pg/g concentrations. First, there is poorer reproducibility in collaborative studies, with mean relative standard deviations generally increasing with lower concentrations of analyte, irrespective of analyte or analytical method.[1] Second, there is poor interlaboratory agreement in round-robin studies, with results varying by orders of magnitude. For example, for a single sample of milk powder, distributed by the International Atomic Energy Agency, reported lead levels varied from 17 ng/g to 247 μg/g, over 5 orders of magnitude.[2] For marine biological samples, an interlaboratory study conducted by the National Research Council of Canada also found widely varying lead levels.[3] For swordfish muscle, the material with the lowest lead level (100 ng/g by isotope-dilution mass spectrometry), 41 laboratories reported levels ranging from less than 50 ng/g to 60.5 μg/g. Of the 23 laboratories which submitted quantitative results, only 13 remained after rejection of outliers. These data indicate the absence of, or failure to use, quality control measures which can detect orders of magnitude errors at low analyte concentrations.

Although numerous quality control protocols exist for validation of results and methods, most cover the commonly known measures, such as recovery studies, blank control, analytical environment, the use of standard reference materials with certified levels of analyte (SRM's) or laboratory reference materials (LRM's), application of alternative methods of analysis, and collaborative and round-robin studies. Few quality control procedures have dealt with the peculiarities of microtrace analysis: adventitious contamination and reagent blanks, and reporting concentrations which are less than the detection limit. This presentation will discuss measures of quality control which are new or have been ignored in analytical practice. At this point, they are presented with a view to encouraging discussion.

2   SAMPLE WEIGHT TEST

Background

Recovery studies are the most frequently used method of sample result or method validation.   They do not reflect analytical accuracy, however, because they only evaluate recovery of the analyte added to the sample and tell us nothing about the amount of analyte present in the sample.   That is, they give no indication about the accuracy of the unspiked sample signal.   Thus, recovery can be 100%, yet analytical results can be orders of magnitude in error.

Situations causing such errors are contamination of samples but not blanks, contamination of blanks but not samples, uncorrected background in atomic absorption spectrometry contributing to a portion or to all of the analyte signal, or invalid baseline in chromatography or stripping voltammetry.

The existing ways to evaluate the presence of any of the above errors are to include appropriate RM's with baseline-levels of analyte, or to analyse the samples using a completely independent method of analysis.   The former is infrequently used because of availability and cost, and even when RM's are used, analysts prefer to choose those with higher concentrations because the quality control results "look better".   The use of an independent method of analysis is usually impractical due to productivity demands on the analyst.

This section describes a test, based on varying the sample weight, which overcomes the above problems with little additional time or cost to the analyst.   The authors have not seen the sample weight test described previously in the literature or in various quality control manuals.

Procedure

The sample weight test involves analysing 2 different weights of the same sample.   One weight should be at least twice that of the other.   The number of replicates determined at each sample weight depends on the critical nature of the sample, the homogeneity of the sample and the precision of the method.

If the same analyte concentration is obtained for the 2 sample weights, then the result can be considered accurate.   If different concentrations are obtained for the 2 sample weights, then the results should not be reported and a cause for the discrepancy should be sought.

The test should be applied to all samples of a critical nature (sample result validation) and to all test samples when a method is being validated.

<u>Example</u>

A simple example of the test is the analysis of seawater (National Research Council of Canada SRM, CASS-1) certified for lead at 0.25 ng/g. The analytical method consisted of adding 10 mL perchloric acid to seawater in a 250-mL Pyrex Erlenmeyer flask, evaporating the seawater to perchloric acid fumes, precipitating the lead, and analysing the dissolved precipitate by graphite-furnace atomic absorption spectrometry (GFAAS). Two sample weights of seawater were tested, 100 g and 150 g. (One sample weight was not twice the other, but the difference was sufficient to illustrate the effect.) When evaporations were carried out in Pyrex flasks, 0.4 ng/g was found for the 100-g sample weight and 0.6 ng/g was obtained for the 150-g sample weight.

The discrepancy in concentrations for the 2 sample weights revealed an error in the method, substantiated by poor agreement with the certified concentration, and an explanation was sought. Although the treatment of the samples and reagents used were identical, the problem was thought to be associated with flask contamination. When high-purity quartz flasks were used instead of Pyrex, the concentrations obtained for the 100-g and 150-g sample weights were, within experimental error, equal (0.25 and 0.22 ng/g), and agreed with the certified level. Thus, the samples were becoming more contaminated than the reagent blank when Pyrex flasks were used.

The explanation of the effect, based on a subsequent test using deionised water, is that acidic solutions do not leach lead from Pyrex, while neutral or basic solutions leach detectable quantities. During evaporation of water from the acidified samples, a portion of water vapour was condensing on the walls. The condensed water was neutral, leaching lead as it returned to the bulk solution. The more water present in the flask, the longer the leaching time and the greater the contamination. Reagent blanks, (containing the same amount of acid as the samples,) did not leach any detectable amount of lead because they contained no added water. The problem was not observed for the quartz flasks which contain less lead than Pyrex.

Thus, the test was able to identify an analytical problem which would not have been detected using recovery studies.

3   STANDARD AND LABORATORY REFERENCE MATERIALS

It is well known that SRM's and LRM's should be chosen for their proximity to actual samples both in sample matrix and analyte concentration. Few quality control manuals, however, suggest the use of SRM's or LRM's to test for non-routine analytical errors.

Many methods perform well until unusually difficult samples are encountered or instrumentation is improperly adjusted. For example, in GFAAS, deuterium background

correctors require readjustment each time a lamp is
changed. There is no way, however, to guarantee that the
adjustment made is optimum, just as there is no way to
guarantee that high levels of non-specific absorbance are
being adequately corrected. Incompletely corrected
background can cause positive or negative absorbance
readings when no analyte is present. The effect is
sample-specific, so normal use of SRM's to monitor
accuracy will not reveal that a problem exists. To
overcome this potential error, each analytical batch can
include a RM or synthetic sample solution containing a
low analyte concentration in a matrix which produces high
background. Good accuracy obtained for this solution
would routinely confirm a properly adjusted background
corrector.

There has been little emphasis in quality control
manuals to assure that one of the RM's used has analyte
at a level near the detection limit of the method. This
can avoid major errors due to invalid blanks, i.e.,
situations in which the contribution of contamination to
samples and blanks differs in spite of seemingly rigorous
identical treatment of both.

4   DETECTION LIMITS AND BLANKS

Background

The detection limit of a method is usually defined
as 3 times the standard deviation (3s) of the blank or
replicate low-level sample measurements. It is usually
determined by running 10-20 replicate blanks and using
these to calculate the standard deviation. The problem
with this approach in microtrace analysis is that it does
not take into consideration the variables which can
affect the detection limit on a day-to-day basis:
adventitious contamination, cross-contamination from
actual samples, analyst expertise, laboratory, analyte
level in the reagent blank, and instrument optimisation.
Thus, the detection limit will often vary on a day-to-day
(batch-to-batch) basis, and will have little relationship
to that obtained initially by running 10-20 blanks.

Procedure

One way to overcome the above problem is to run
multiple blanks within each analytical batch. Indeed,
quality assurance manuals should require that multiple
blanks be included in each analytical batch. The number
of blanks to include in each batch will obviously affect
analytical productivity, so guidelines, grouped according
to the purpose of the analysis, are proposed (Table 1).

In summary, three detection limits are available to
the analyst depending on the method of calculation. The
method detection limit, obtained when the method is first
tested by running 10-20 replicate blanks, is valid only
on the day of the test. It can be used most beneficially

Table 1. Guidelines for choosing the number of blanks to include within each batch.

Analytical requirements	Guideline
The concentration of each sample is of importance irrespective of whether the analyte level found is high or low, e.g., for monitoring the quality of a sample, or for epidemiological studies.	A close estimate of the true detection limit for each analytical batch is needed. Statistically, the only way this can be obtained is to run 4-8 blanks within the batch. If adventitious contamination is a potential problem, then 6-8 blanks must be used.
The concentration of each sample is of importance, but only at concentrations well above the detection limit of the method, e.g., for compliance work or contaminant screening.	A minimum of two blanks are needed; however, more must be included when the extent of adventitious contamination is unknown.
Surveys of large numbers of samples where individual sample concentrations are only needed for estimating means, medians and ranges, e.g., dietary intake studies.	The number of blanks needed in each batch depends on the number of analytical batches run in the survey. Specifically aiming for a minimum of 9 degrees of freedom, 2 blanks in each batch are needed if 10 or more batches are included in the survey. The detection limit can be estimated from the pooled (over all batches) standard deviation of the blanks, and will represent a survey detection limit. Individual batch detection limits will not be valid. If only 2 batches are included in the survey, then 5-6 blanks should be included in each.

to confirm the use of high-purity reagents and good blank control by the analyst. The batch detection limit, obtained by running 4-8 blanks in each analytical batch, is valid for each sample in each batch. The survey detection limit, obtained by running 2-3 blanks in each batch, is valid for the total survey but should not be used for critical samples within a batch.

Each analytical batch, irrespective of the purpose of the analysis, should contain a minimum of 2 reagent blanks. This cannot be overemphasized, because there is a tendency among those with clean rooms to assume that contamination is negligible and consistent, and that one blank should suffice. A similar tendency exists among mass spectrometrists because of the good sensitivity of the instrumentation. In particular, isotope dilution mass spectrometry, regarded as a definitive technique, is just as capable of producing inaccurate results as less expensive instrumentation. Its reputation as a definitive technique is due more to the expertise of the analytical chemists using it than to its ability to compensate for losses during analytical work-up.

5 MULTIPLE BLANKS AND SAMPLES AND DETECTION LIMITS

Background

Although the analysis of multiple blanks is needed
to assess the detection limit of a method, once it has
been calculated, the inclusion of multiple blanks and/or
samples in an analytical batch actually reduces the
detection limit.   The reason for this is that the
statistical confidence associated with the 3s definition
of the detection limit refers to the situation when a
single blank signal is subtracted from the sample-plus-
blank signal for a single sample.

Procedure

Table 2 illustrates how the detection limit is
affected by multiple blanks and samples.   Furthermore,
the reduction is effected by increasing the number of
blanks, the number of sample replicates, or both.   For
example, by increasing the number of blanks in a batch to
5 and the number of sample replicates to 3, then the
effective detection limit can, with the same degree of
confidence, be reduced from 3 to 1.55.

Table 2. Influence of the number of blanks and number
          of sample replicates on the effective
          detection limit.[a]

Number of blanks	Number of sample replicates and effective detection limit			
	1	2	3	4
1	3.00	2.60	2.45	2.42
3	2.45	1.94	1.73	1.62
5	2.32	1.77	1.55	1.42

[a] The assumption is made that the method has a true 3s
detection limit of 3.

6 BLANK OUTLIERS

Background

As more laboratories include multiple blanks in each
batch, the observation of blank outliers will become
commonplace.   For example, in a survey of lead in human
milk,[4] blank statistical outliers due to adventitious
contamination were frequently observed.   This section
discusses the circumstances under which the outliers,

BATCH 1

BATCH 2

BLANK	SAMPLES		BLANK	SAMPLES	
SIGNALS	+ BLANK		SIGNALS	+ BLANK	
MEAN = 1	MEAN = 2		MEAN = 1.3	MEAN = 2.3	

SURVEY RESULTS	BATCH 1	BATCH 2 (OUTLIER INCLUDED)	BATCH 2 (OUTLIER REJECTED)
SAMPLE AVERAGE	1.0	1.0	1.3
SAMPLE 1	1.0	0.7	1.0
SAMPLE 2	1.3	1.0	1.3
SAMPLE 3	0.8	0.5	0.8
SAMPLE 4	0.7	0.4	0.7
SAMPLE 5	1.2	2.7	3.0
SAMPLE 6	1.0	0.7	1.0

<u>Figure 1</u>. Simulation of blank and sample signals. Signals in batch 2 are equal to those in batch 1 except for blank 5 and sample 5 (batch 2), which have been "adventitiously contaminated" equally.

prior to subtraction from sample solution concentrations, should be retained or rejected.

<u>Procedure</u>

Figure 1 illustrates the effect of blank outlier rejection on individual sample concentrations as well as on the mean sample concentrations. In each batch, 6 blanks and 6 samples are run. The signals of blank and sample solutions are identical in both batches except for 1 blank outlier and 1 sample contaminated at the same level as the blank outlier in batch 2. In batch 1, the mean signals of the blanks and samples are 1.0 and 2.0, respectively. In batch 2, the mean signals as a result of contamination are 1.3 and 2.3, with 0.3 representing the average contribution from the added contamination. Sample concentrations are calculated by averaging the blank signals, and subtracting the mean from the individual sample signals.

In both batches, the average sample concentration is equal to unity if the blank outlier is retained. In batch 2, however, rejection of the blank outlier will result in an erroneous sample mean of 1.3 because, in practice, it would be impossible to identify and reject the contaminated sample. Thus, when estimating mean concentrations for surveys, blank outliers must be included.

If, instead of the average concentration, one examines the concentrations of the individual samples, one notices that best between-batch agreement for the individual samples is obtained when the blank outlier is rejected. The only sample not agreeing between batch 1 and 2 (Figure 1, column 1 vs column 3) is that of sample 5 which had been "contaminated"; i.e., 5 of the 6 sample results agree after blank rejection. If, on the other hand, the blank outlier had been retained, poor agreement for all 6 of the sample concentrations (column 2 vs column 1) would have resulted.

Thus, when it is desirable to make comparisons among individual samples (e.g., for epidemiological studies correlating concentrations with attributes of individual samples), blank outliers must be rejected. In effect, this reduces between-batch bias.

Irrespective of the application of the data, blank outliers must be included when estimating the detection limit.

In conclusion, the method chosen for treating blank outliers will depend on the application of the results. For those studies in which both means and individual sample concentrations are important, both methods of calculating results will be required for each analytical batch.

7   BLANK DISTRIBUTION WITHIN THE ANALYTICAL BATCH

Background

When running multiple reagent blanks, there is a tendency, even within our own laboratory, to group all the blanks together within the analytical batch. This can lead to a serious underestimation of the detection limit.

Procedure

If possible, blanks should be distributed evenly throughout the analytical batch. Alternatively, the blanks can be grouped at the beginning at the run and reanalysed at the end of the run. In this case, however, the standard deviations for the groups at the beginning and end of the run would be calculated individually and pooled. A third approach is illustrated in the example below.

Example

In the analysis of aluminum in infant formulas,[5] only the samples were digested. The standards were not digested and could be used over several batches. As a result, the analytical method included one reagent blank for the standards, and 3 separate digested reagent blanks for the samples. The standards (including the single blank) were run at the beginning and end of each analytical batch. The 3 sample blanks, were run sequentially, immediately following the first standard set, and 26 samples were run after the sample blanks.

Table 3 shows how the precision for the standard and sample blanks varied in 3 analytical batches. In the

Table 3. Effect of blank positioning within the batch on blank standard deviations for aluminum (ng/mL).

Solution	Batch 1	Batch 2	Batch 3
Standard blank (single solution run before and after samples)	0.61	0.56	0.36
Sample blanks (3 individually digested solutions run sequentially within batch)	0.21	0.18	3.40

first 2 batches, the precision for the single standard blank was significantly poorer than the precision for the 3 sample blanks. The reason for the poorer precision for the standard blank solution was that there was an unidentified instrument artifact which appeared to gradually shift the baseline over the progression of the run. In batch 3, the poorer precision for the sample blanks was due to contamination of one of the solutions.

To overcome the differences in blank precision and to avoid underestimation of the detection limit of the method, the standard deviations for the standard and sample blanks were pooled for each batch. An alternative and more correct approach would have been to evenly distribute the sample blanks among the 26 samples. Pooling the standard deviations would still be required.

8 REPORTING SURVEY DATA BELOW THE DETECTION LIMIT

Background

In some analytical surveys, a significant portion of results falls below the detection limit. In the past, the detection limit, zero or one-half the detection limit have been most frequently used to represent concentrations less than the detection limit; however, all these are arbitrary and can result in significant bias when estimating mean concentrations or when using the results for other calculations, such as dietary intake estimations. This section proposes an alternative method of reporting such results.

Figure 2 illustrates 1000 computer generated points with a gaussian distribution, a theoretical mean of zero and standard deviation of one. These points can represent results for 1000 samples, each having an analyte concentration of zero. The standard deviation of the method can be considered to be unity. Figure 3 represents 3 distributions, one with a true (population) mean of zero (using the data in figure 2), one with a true mean of 0.5, and one with a true mean of 3 (i.e., the detection limit of the method). For the latter distribution, although sample concentrations equal the detection limit, half of the concentrations will be less than the detection limit and some less than zero. Thus,

CONC.

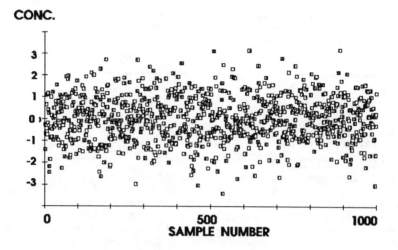

SAMPLE NUMBER

Figure 2. Computer-generated representation of the
analytical results obtained assuming all
samples contain no analyte and the standard
deviation of the method is unity. Note that
about half of the concentrations will be less
than zero.

irrespective of the circumstance, if we report those
concentrations which are less than the detection limit as
zero or as the detection limit, the mean will be biased.
    To overcome any potential bias, it is proposed to
report results as calculated, including concentrations
less than the detection limit and less than zero.

Procedure

    It is well known that for multiple data points with
a gaussian distribution the Student t-test can be used to
estimate, with a specified level of confidence, the
interval within which the true (population) mean can be
found:

$$\mu = X \pm \frac{ts}{\sqrt{n}}$$

where μ is the true mean, X is the actual (sample) mean,
s is the calculated standard deviation of the results, n
is the number of degrees of freedom and t, Student's t,
is a statistical parameter which depends on the desired
degree of confidence and the value of n.
    Figure 4 illustrates application of the t-test
(99.9% confidence) to the first 2 distributions in Figure
3.   It is clear that even at this high level of
confidence, it is possible to distinguish samples with a
true mean of 0.5 from zero. This is valid, however, only

PROBABILITY

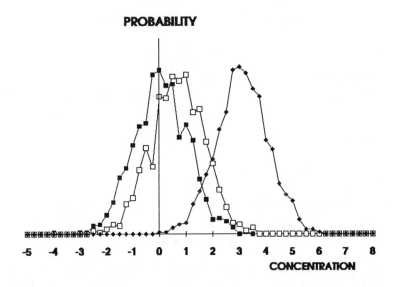

CONCENTRATION

Figure 3. Probability distribution of 3 populations, each with a theoretical standard deviation of unity. The theoretical means of the distributions are, from left to right, 0, 0.5 and 3.0.

when results are reported as calculated.

Thus, when a significant number of survey results fall below the detection limit, all results should be reported as calculated, including those less than the detection limit and less than zero. The distribution of the results should then be evaluated for statistical distribution and, if necessary, converted to a gaussian distribution using a transformation. Consulting a statistician is highly recommended for this purpose. Next, the mean and confidence interval of the mean can be calculated. If the lower confidence interval of the mean is greater than zero, then the mean can be published even though it may be below the detection limit of the method. Also, if the mean is distinguishable from zero, then in all likelihood, the median can also be estimated using these data.

It should be noted that this approach is only valid with large numbers of survey data, and that the number of degrees of freedom is not the number of samples minus one, but is a complex function of the numbers of analytical batches, reagent blanks, and sample replicates run within each batch. Also, the validity of the approach requires the generation of unbiased analytical results near zero concentrations. This is the responsibility of the analyst, and tests are needed to confirm this.

Finally, the approach cannot be used when reporting individual sample concentrations which are less than the detection limit. Such data must be reported as being

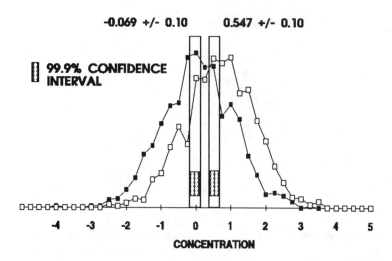

-0.069 +/- 0.10   0.547 +/- 0.10

<u>Figure 4</u>. 99.9% confidence intervals of means for 2
populations, each with a theoretical standard
deviation of unity and means of 0 and 0.5.

less than the numerical value of the detection limit
because little statistical confidence can be placed in a
single numerical value of a concentration which is near
the detection limit.   Similarly, ranges must also be
represented in this manner because they reflect
individual sample concentrations.

9   LABWARE SEGREGATION AND RANDOMISATION

<u>Background</u>

It is well known that containers used to store or
digest concentrated solutions of analyte should not be
subsequently used to hold dilute solutions because of the
potential for carryover.   Thus, the prior analytical
history of labware must be documented, and containers
used for concentrated solutions should be segregated from
those used for dilute solutions.
This concept has, unfortunately, been extended to
the analytical batch level, where vessels have been
segregated as to whether they were used for blanks or
samples, and, in isotope dilution mass spectrometry, by
whether or not isotopes had been added when analysing
samples or blanks.[6] This practice is erroneous.

<u>Procedure</u>

All labware used in the analytical work-up of
reagent blanks and samples must be randomised prior to
each use. If analyte carryover from prior use of the

vessels is a problem, then it will exist whether or not the labware is randomised. However, through randomisation and the inclusion of multiple blanks, analyte carryover can be identified. If vessels are segregated by blanks and samples, the blanks will always look good, but the samples may be contaminated.

Similarly, concentrated solutions of isotopes should be stored in segregated containers. However, once the isotopes enter the analytical process, then the labware used must be completely randomised, and the potential for isotope carryover checked through the use of multiple blank analyses.

As an extension to randomisation of labware, blanks and samples should be interspersed during actual performance of the analysis.

REFERENCES

1. K.W. Boyer, W. Horwitz and R. Albert, Anal. Chem., 1985, 57, 454-459.
2. R. Dybczynski, A. Veglia and O. Suschny, "Report on the intercomparison run A-11 for the determination of inorganic constituents of milk powder," IAEA/RL/68, International Atomic Energy Agency, Vienna, Austria, 1980.
3. S.S. Berman and V. Boyko, "First intercomparison exercise for trace metals in marine biological tissues, NRC BT1/TM," NRCC No. 25324, National Research Council of Canada, Ottawa, Canada, 1985.
4. R.W. Dabeka, K.F. Karpinski, A.D. McKenzie and C.D. Bajdik, Food Chem. Toxicol., 1986, 24, 913-921.
5. R.W. Dabeka and A.D. McKenzie, Food Addit. Contam., 1990, 7, 275-282.
6. L.A. Machlan, J.W. Gramlich, T.J. Murphy and I.L. Barnes, I.L. "The accurate determination of lead in biological and environmental samples by isotope dilution mass spectrometry, "Accuracy in Trace Analysis: Sampling, Sample Handling, Analysis", U.S. Department of Commerce, Gaithersburg, MD, Vol. II, NBS Spec. Pub. 422, 1976, pp 929-935.

Experience with the IUPAC-1987 Harmonized Protocol for Method-Performance Studies: Suggestions for Revision and Application to Internal Quality Control Systems

William Horwitz and Richard Albert

US FOOD AND DRUG ADMINISTRATION, WASHINGTON, DC 20204, USA

Five years of experience with the IUPAC-1987 Harmonized Protocol for the design, conduct, and interpretation of method-performance studies suggest three substantive, but minor, revisions: (<u>1</u>) Remove the inefficient double split-level design to avoid misinterpretation of results; (<u>2</u>) clarify the explanation of the term "material"; and (<u>3</u>) change the outlier-removal procedure to a 2% (2-tail) probability level from 1% to remove additional aberrant values. The empirical formula developed for estimating the <u>among-laboratories</u> relative standard deviation (RSD_R, %) as a function of concentration, C, expressed as a decimal fraction,

$$RSD_R = 2^{(1 - 0.5 \log C)} \approx 2C^{(-0.1505)},$$

can be applied to the within-laboratory internal quality control situation by using the within-laboratory relative standard deviation, RSD_r, obtained by multiplying RSD_R by a factor of 1/2, thus:

$$RSD_r = 0.5 \times 2^{(1 - 0.5 \log C)} = 2^{(- 0.5 \log C)} \approx C^{(-0.1505)}.$$

The formula for RSD_R predicts that starting with a concentration of 100% (pure materials; C = 1.00), RSD_R will be 2%, increasing by a factor of 2 for each decrease in C of 2 orders of magnitude. For parts-per-million concentrations (C = 10^{-6}), RSD_R = 16%; within-laboratory variability, RSD_r, is roughly one-half of RSD_R. This formula, based upon the examination of the precision from almost 7000 interlaboratory data sets published over a period of about 100 years, has been found to be a useful guide to acceptable precision in the fields of agricultural, nutritional, pharmaceutical, and geochemical analysis. The predicted values provide analysts with a rational reference point for interpreting both the within-laboratory and among-laboratories uncertainty of the analytical portion of the total variability of concentration estimates. RSD_R is more or less independent of analyte, method, and matrix, in the absence of more significant information. However, at the trace levels of current interest (ppm, ppb), where the RSD_r routinely is 10-50% and RSD_R is 20-100%, several hundred independent measurements would be required to identify a statistically significant alternative distribution. Therefore, it is suggested that, at least initially, a standard deviation based on the above empirical formula be used in quality control charts relying on a "z-score," defined as the deviation of a value (e.g., from a laboratory) from a reference value, all divided by a concentration-dependent standard deviation.

Ten years of correspondence, small working group meetings, and three general conferences were required to develop the initial IUPAC "Protocol for the Design, Conduct, and Interpretation of Collaborative Studies".[1] The history of this effort has been chronicled by Horwitz.[2] Organizations promulgating standard methods of chemical analysis that have now adopted this protocol are AOAC International, American Oil Chemists Society (AOCS), International Association for Cereal Science and Technology (ICC), International Commission for Uniform Methods of Sugar Analysis (ICUMSA), and Nordic Analytical Committee (NMLK). There may be others that we are not aware of. The one major organization that supplies methods of analysis to the Codex Alimentarius that has not formally adopted it is the International Organization for Standardization (ISO), but their primary document, ISO 5725-1986, now in the final stages of revision, uses the same basic requirements as the IUPAC protocol, namely, 1-way analysis of variance, and the Cochran-Grubbs outlier removal treatment at the 1% probability level. The only difference that may occasionally lead to reporting different statistical performance parameters that we have noted thus far is the failure of the ISO document to specifically provide for recycling in the search for outlying values. This is not a fatal flaw because a well-conducted method-performance study should not have any unexplained outliers; explained outliers are removed before any statistical treatment or outlier search is initiated.

We have had five years of experience with the IUPAC-1987 protocol; consequently we are now in a position to recommend reconsideration of some of its provisions. Based upon our experience in examining the published record of reliability of interlaboratory studies, we are also able to make some suggestions with regard to internal quality assurance schemes.

1   BACKGROUND

For many years, we have been studying acceptable performance in analytical chemistry, primarily with respect to precision and with respect to what ISO is now calling "trueness." There is no controversy with respect to defining precision -- how close values in a set are to each other. Scientists in general, and analytical chemists in particular, know that they must always indicate which precision they mean -- within run, within batch, within days, and within laboratory, and among laboratories, methods, and matrices. But considerable misunderstanding has accumulated with respect to "accuracy" because of failure to use an appropriate modifier. The revised version of ISO 5725 is attempting to standardize the concepts and nomenclature in the following way:

Accuracy is used in a generic sense of overall error of a single reported value -- as the difference of an <u>individual, measured value</u> from the true, correct, or assumed value. It is composed of both a trueness term and a precision term.

Trueness is used in the sense of accuracy of the <u>mean</u> -- the difference of the <u>average of a finite number of individual values</u> from the true, correct, or assumed value.

Bias is used in the sense of accuracy of the <u>long-term average</u> (expectation; expected value) of a series of averages -- the difference in the expected value (theoretically equal to the average of an infinite number of independent, individual values) from the true, correct, or assumed value. The major bias in analytical chemistry is the laboratory bias, but there also can be analyst (within-laboratory) bias, method bias, and matrix bias (due to interference). Method bias is the average of a large number (theoretically an infinite number) of individual laboratory biases, and the variance of these individual laboratory biases is denoted as σ_L^2 in ISO 5725.

There are important conceptual differences among the three terms -- accuracy, trueness, and bias -- but for practical purposes in analytical chemistry and for large numbers of values, trueness and bias are close together, but accuracy of an individual value can be quite variable because it does not have the moderating effect of its neighbors on averages. Therefore, no revision is necessary in the IUPAC-1987 protocol except possibly some changes in nomenclature to ensure harmonization with the revised ISO 5725, which is in conformity with good statistical practices.

Several years ago, the Codex Committee on Methods of Analysis and Sampling of the Joint FAO/WHO Codex Alimentarius program for the establishment of international food standards developed a check list of items required to evaluate methods of analysis for adoption as Codex Methods. The Committee, at its 18th meeting in November 1992, undertook to revise its checklist to conform with the IUPAC-1987 protocol. Along the way the Committee had the benefit of several years of experience with the application of the protocol to interlaboratory studies. It made several recommendations for consideration by the IUPAC Working Group in charge of harmonization activities. The two most important recommendations, aside from purely editorial and grammatical changes, are (1) the elimination of the double split level from the permitted designs for estimating the statistical parameters of methods, and (2) an amplification of the definition and explanation of the use of the term "materials." Furthermore, based upon the published record, we recommend a change in the outlier-removal treatment which is critical to the reporting of the "best estimate" of the statistical parameters resulting from an interlaboratory study.

2   PROPOSED CHANGES

Elimination of the Double Split-Level Design

Although some doubt was expressed at the harmonization meeting held in Geneva in 1987 as to the validity of the double split-level design (a closely matched pair of materials, with each of the individuals analyzed as unknown ("blind") duplicates for a total of 4 analyses per level; statistically a split level is a single level), this design was accepted as one of the alternatives for estimating statistical parameters from method-performance studies. No change is recommended in the ordinary split-level (Youden pair) design, which consists of a material split into two parts which are so close in composition that they can be considered as having the same within-laboratory and among-laboratories variance. Usually this is accomplished by adding a small amount of analyte or diluent to half of the material and analyzing each of the slightly different levels once. Further statistical review shows that the results from the design, which requires the use of 2-way analysis of variance (ANOVA) for the statistical analysis, cannot be correctly interpreted. This type of ANOVA is now seen to be inconsistent with the requirement for the 1-way ANOVA specified elsewhere in the IUPAC-1987 protocol, as well as in ISO 5725. Furthermore, an interaction term that appears in a typical 2-way ANOVA depends on the arbitrary choice of the spacing of the sublevels of the split levels. Such an interaction term, when significant, is physically uninterpretable. In other words, the estimate of the statistical parameters developed would depend on which concentrations happened to have been chosen for the method-performance study, conditions which cannot be tolerated in the production of a "standard" protocol. Another problem which also applies to the single split-level design is that a rigorous definition of "closely matched" is not provided. Dr. H.C. Hamaker, who is considered the "father of ISO 5725," in a personal communication to Dr. Horwitz of February 3, 1987, stated, ". . . the formula $\sigma_R^2 = \sigma_L^2 + \sigma_{LP}^2 + \sigma_r^2$ on which [the double split-level design of] NEN 6303 is based, is fundamentally wrong." Both Dr. Mandel and Dr. Wilrich of ISO TC 69 on

Statistics have objected to its use. The design is legitimate; it is the interpretation in specific cases that is impossible.

In addition, a more practical consideration is that the double split level is very wasteful of resources. It emphasizes repeatability (within-laboratory variability) at the expense of the far more important reproducibility (among-laboratories variability). Repeatability can be developed inexpensively as part of the internal quality control procedures of each laboratory; reproducibility can be obtained only as part of a well-planned, organized, expensive effort on the part of many volunteer laboratories. This practical objection is equally applicable to those protocols that specify the use of 5 replicates per laboratory, and more if outlying results appear. The design is not inappropriate; but the analytical time of these expensive interlaboratory studies can be more efficiently spent in analyzing additional matrices or levels as unknowns.

Therefore the <u>double</u>-split level design should be removed from the harmonized protocol.

Amplification of the Explanation of the Term "Material"

The term "material" is a synonym for what the chemist colloquially has called "sample" -- a specific matrix/analyte/concentration combination. A change is recommended because use of the word "sample" for this concept resulted in misinterpretations and ambiguities.[3] The principle behind the use of the term "material" is to encompass the range of matrices to which the method is applicable. Much of the work involving applicability should be done during the development phase of the method to show which commodities can be handled by the method and which cannot. Therefore, it is recommended that the design section of the protocol be amplified to relate number of materials directly to applicability of the method. The statement recommended by the attachment to the Codex document is,

> "This attribute defines the applicability of the method. For application to a number of different commodities, a sufficient number of matrices and levels should be chosen to include potential interferences and the concentrations of typical use."

The elimination of the double split-level design removes one of the other semantic problems with this term as well -- whether a [double] split level involves 1, 2, or [4] "materials." [The statistical answer in all cases is 1 because the concentration levels are not distinctly different.]

Outlier Treatment

At the meeting which approved the Harmonized IUPAC-1987 protocol, reservations were expressed by some participants that the outlier-removal treatment adopted was not necessarily the most appropriate. This was acknowledged by the attendees, but the procedure adopted (Cochran, Grubbs, and multiple Grubbs at the 1%, 2-tail probability level (i.e., 0.5% at each tail), applied in the order given) was accepted as a reasonable compromise, based upon the ISO 5725-1986 model. Subsequently, ISO TC 69 replaced their Dixon test, which was in use at the time, with the Grubbs test. The Dixon test was a remnant of the slide rule era, which gave way to modern computer computations. However, the reservation remained that the outlier tests chosen, although reasonable and easy to apply, were not always the "best." The thought was expressed that a judicious case-by-case approach was far better. In practice, of course, this is what is done, but a free choice of outlier-removal procedures is not conducive to harmonization and consequently could not be adopted.

As we expanded our review of the precision parameters of methods of analysis used for regulatory purposes from the initial drug commodity area to dairy products, foods, pesticide formulations, and even biological and geological reference materials, we found that the IUPAC-1987 outlier treatment had to be abandoned occasionally because it failed to remove obvious "sore-thumb" outliers, particularly when only a few laboratories (≤ 10) were involved. When we investigated these situations further, such values were often found just inside the cutoff 1% probability level. Such borderline decisions, however, can be expected with any system that checks against a specification value.

However, the most definitive examination of the performance of the IUPAC-1987 outlier-removal treatment is given in our investigation of the reported values for the NIST Standard Reference Materials (SRMs).[4] We reviewed the extensive compilation of Gladney et al.[5] This compilation of reported literature values from research, method development, and quality control papers provides, in our opinion, the best performance that the analytical chemistry profession has to offer because it displays uncensored values that are open to public exposure and potential criticism. Here we have an assigned "true value," and experience shows that analysts obtain better precision when they know the answer than when they are analyzing unknowns. Therefore, the variability exhibited by this population of analytical values can be subjected to various outlier treatments. The treatment that provides the consistently "best" estimate of the mean and uncertainty can be accepted as a reasonable way to remove outliers. "Best" in this sense is providing a mean value closest to the true value, but yet not removing an

excessive number of outliers. These are somewhat
conflicting requirements so we can never say their
application will be the "best" outlier treatment. In the
region of low concentrations the inherent variability is
so large that it would take a statistical analysis of an
unreasonable number of values (i.e., thousands) to
determine if one distribution is significantly different
from another, e.g., if a population is normally or
nonnormally distributed in order to apply the "proper"
statistical outlier-removal procedure.

We subjected the data from 11 biologically related
SRMs for 29 elements, where at least 8 quantitative values
were available per analyte/matrix combination (total 117),
to several outlier treatments as follows:

(1)   The IUPAC-1987 procedure which, in this case of
      assuming single values per report (hence no
      Cochran test was applied), consisted of only the
      Grubbs and multiple Grubbs tests as extended
      beyond 40 values by Kelly,[6] recycled until
      values were no longer flagged, with an automatic
      stop at removal of 22.2% (2/9) of the number of
      laboratories.

(2)   The consensus technique, as reported by the
      compilers, who first removed less than 1% of the
      total data as "clearly beyond the limits of
      acceptability." Then all values beyond the
      mean plus or minus two standard deviations, S,
      were removed and the recalculated mean and S
      were reported as the "consensus mean and
      associated standard deviation."

(3)   The NIST values with the reported uncertainty
      assumed to be a standard deviation. Although
      this assumption may not be true, it does result
      in values which are "close enough." The
      resulting variability is not necessarily used as
      a reference point and, therefore, its absolute
      value is unimportant for the present purpose.
      It is only given as an indication of the
      uncertainty ascribed by the certifiers to the
      reference mean.

The resulting means and standard deviations were
recalculated and reexpressed in terms of among-
laboratories relative standard deviations, RSD_R, to permit
a common basis for comparison. These RSD_R values were used
as the numerator in a fraction, designated as HORRAT,
with an RSD_R value calculated from the Horwitz equation[7] as
the denominator. HORRAT has been found to be the most
useful single statistical parameter to indicate the
acceptability of the precision of the reported
interlaboratory concentration values, C, because it
normalizes the effect of the single most important factor

affecting variability of analytical results -- concentration.[8] The C values of the data in this review extended over the extremely wide range of 5 ng/g (ppb) to 4%.

3 RESULTS

The results for the 206 analyte/matrix combinations with certified and informational concentrations with more than 8 values are in the original paper.[4] To simplify the presentation, the data for only the 117 certified analyte sets will be discussed here.

Systematic Error

A comparison the NIST certified values with the consensus values established by the authors of the compilation showed substantial agreement over the entire 7 decades of concentrations covered. Only 6 (5%) consensus means differed by more than 10% from the certified value. These differences came from easily explained sources: (1) from trace elements present at very low levels (C = 10^{-6}-10^{-8}), which present chemical problems (Be, Cr, V), and (2) from the elements that present environmental problems from contamination (Fe, Cr).

When the mean values are calculated by the IUPAC-1987 outlier-removal treatment, however, 10% differ by more than 10% from the certified values, and are well scattered among the analytes. In other words, the IUPAC treatment is more likely to result in mean values deviating from the certified value. However, in general, the deviations are very obvious since approximately 2/3 (1 standard deviation) of the values from both sets (consensus and IUPAC-1987) are within 4% of the certified value.

Random (Analyte/Matrix) Error

The random errors are here measured as the distribution of the 102 HORRAT values from those analyte/matrix combinations with at least 3 certified values. (The averages from 1 or 2 values can be considered as almost meaningless.) The average HORRAT value from the reported consensus data sets is 0.8; the equivalent average HORRAT value developed from the variability calculated by the IUPAC-1987 protocol is 1.6, about twice as large, but nevertheless acceptable. It is hard to imagine that any better evidence can be presented to suggest what can be considered as a suitable outlier treatment -- to attempt to obtain a distribution that centers itself approximately at the assigned reference value with a minimum, reasonable variability. Reasonable variability is defined here as the variability exhibited by the historical record of over 7000 interlaboratory data sets, which approximates a HORRAT of 1. We have specifically published the historical record of almost 1000 data sets from pesticide formulations (Figure 8).[8] Similar precision data plotted as a function of time, awaiting publication, are available from almost 1000 data sets each for mycotoxins and drugs. We anticipate that a detailed examination of the remaining 4000 data sets, from foods, fertilizers, water, minerals, and pesticide and drug residues, will support a conclusion that a modification of the confidence levels for the present outlier-removal treatment (i.e., Cochran, and single and multiple Grubbs) will use the 98% probability level (1% in each tail).

Furthermore, we recommend that organizers of interlaboratory studies should investigate promptly all suspect values as they are received. They should not rely on statistical tests to remove aberrant data. Prompt investigation will often reveal that the method protocol contains ambiguous wording and potential misinterpretations that can result in unwelcome outliers.

Use of Repeatability and Reproducibility Relative Standard Deviations

Too often, organizations utilizing the IUPAC-1987 protocol, or for that matter any protocol, take the results from an interlaboratory method-performance study literally. For example, ISO 5725-1986 advises plotting s_R and s_r as functions of concentration. Instinctively, the points of each variable are connected, often giving a zig-zag appearance. AOAC requires a statement summarizing the s_R, s_r, RSD_R, and RSD_r found, and the mean concentration, or if a number of similar matrices are available over a concentration region, as ranges. The result may be taken as a point function, which is often interpreted as a reference point or as a target value. Thus subsequent appliers of the method will point to their smaller s and RSD values compared to those of the original study to indicate how much better their values are. Such statements are the result of misapplication of statistics. Each reported s value (or its relatives--RSD_R, RSD_r, r, and R) is merely a sample from a population, and a table of values or ranges found merely shows what was obtained in one study. These descriptive parameters are not what is desired from a method-performance study. What is wanted is the answer to the question, "What values of s [or RSD or R] would I find under the same conditions in the future?"

Chemists are familiar with this problem when they are dealing with individual values and averages; this results in the calculation of the confidence interval for means and the confidence interval for individual values. The confidence interval is the region within which future means are expected to fall with 95% (or other value) probability. When applied to individual values, the region is called the prediction interval. Chemists have probably never thought of future values in terms of standard deviations. Here the problem is complicated because it involves the unsymmetrical chi-squared test. But as the number of observations becomes large, the chi-squared distribution does approach the normal so that the final result may be approximated by the Gaussian distribution. Empirically we have found that the region of acceptable values for the standard deviation (and its related parameters) is obtained by using factors of 2 and 1/2 of that predicted by the Horwitz equation. When expressed as HORRAT values ($RSD_{found}/RSD_{predicted}$), the so-called "confidence [or prediction] bounds" are actually 0.5 and 2.0. Despite some skepticism[9] as to the applicability of the general equation to laboratory-proficiency studies, for which the equation was never recommended, the equation remains as an increasingly useful generalization of interlaboratory variability.[10] Practitioners should be warned not to extrapolate the results from a single study to future expectations, which is what is unconsciously being done.

Concentration is the major factor affecting variability, with method, matrix, and analyte as minor factors whose influence should be removed by suitable analytical chemical research. For example, if substantial "error" (as precision, trueness, bias, or any combination of them) persists in going from matrix to matrix, interference effects have not been removed. Performance does improve with time also, but only down to a minimum from which it often deteriorates.[11] Performance also deteriorates when a group of laboratories operating under a well-defined quality control program is released from the program.[12]

In view of the relatively large variability encountered in estimating the precision of method performance at low concentrations, it appears that the Horwitz equation is about the best available indicator of acceptability required for harmonization.[10] Methods that show performance parameters of less than 2 times the precision calculated from the working concentration can be accepted as reliable on the basis of precision. Historically, such an index has tracked the record of AOAC and other performance studies for over a century of analytical work. When sufficient interlaboratory studies of a single method, e.g., Kjeldahl nitrogen or the CB method for aflatoxin in peanuts, have been performed, the pooled RSD_R and associated prediction interval may be substituted for the general term. Until such a degree of confidence is reached, however, the pooled data based upon the historical record are a far better predictor of analytical performance than the results from a single or few performance studies.

4   INTERNAL QUALITY CONTROL

Most analytical work is performed within a single laboratory. In many cases, the method used is an abbreviated version of a standardized, validated method. A reader is alerted to such modifications by such phrases as "essentially method XYZ," or "modified XYZ method" with little detail as to the extent of the modifications and usually no validation data. As long as such a method serves an internal purpose of quality control of production there is no objection to its continued use. But parameters developed with its use are not part of the population of data obtained by the use of the legitimate method since changes in method parameters, especially time, will change systematic or random errors or both.

For the analytical control of repetitive operations, such as the production of a composite food product manufactured to internal, contract, or legal specifications, there is usually no problem in setting acceptable limits. Often charts of the final result, e.g., a single measurement for a generic constituent such as acidity or moisture, plotted as a function of time, is sufficient to control not only the manufacturing operation, but the sampling and analytical operations as well. When a possibility of within-batch as well as among-batches variability exists, the addition of a range chart of several replicates per batch is sufficient to maintain quality control.

A problem develops in the nonroutine laboratory where occasional isolated test samples are the norm and not the exception. Often there is no previous history or experience with regard to acceptable performance parameters. The first line of protection is to automatically perform duplicates and the second is to determine linearity with different test portion weights. Reference materials are unavailable for almost all types of analyses, except for metallic elements. In such cases, the Horwitz equation can provide at least an indication of what might be considered acceptable precision. It can say nothing about bias.

An empirical observation during the formulation of the Horwitz curve that can assist in determining acceptable precision within laboratories is that RSD_r is approximately 1/2 to 2/3 of RSD_R. Consequently, the maximum acceptable within-laboratory RSD_r is the typical RSD_R calculated from the Horwitz equation itself; typical values are approximately half the curve. However, we do not have much confidence in this parameter as a general index because RSD_r is more variable than RSD_R. For methods that are highly individualistic, like counting mold or insect parts or visual matching of spots in thin layer chromatography, the ratio of RSD_r/RSD_R is relatively low, i.e., 0.2-0.4. Analysts can check themselves very well but they cannot check another analyst. Consecutive aspirations of the same solution into an atomic absorption spectrometer provide the within-instrument RSD_r of 1-3% that is often reported as the precision of these methods. If the RSD_r within a laboratory is really wanted, it should be based upon the use of different calibration curves, depending on the frequency of recalibration. At a minimum, RSD_r should be based upon measurements performed on different days.

When an analysis is conducted near its limit of determination, the ratio of RSD_r/RSD_R is relatively high and the within-laboratory variability is so high that it often approaches and occasionally exceeds the among-laboratories variability. For the standard, run-of-the-mill laboratory procedures away from the limits, the Horwitz curve used as a maximum will crudely and conservatively approximate the within-laboratory precision. However, if the laboratory has a historical record of the precision of a specific method, that database can be substituted for the general equation.

A word of caution about the development of a general, individual within-laboratory RSD_r: Performing duplicates simultaneously will underestimate the within-laboratory precision. What is desired is the precision that can be used with future analyses, and this requires the use of at least different-day measurements. The frequent advice to determine the standard deviation of the blank from 10 measurements falls in the same category. Literally, analysts will take 10 consecutive measurements, which will grossly underestimate the variability of the blank, particularly if the blank is sensitive to fluctuations in temperature, barometric pressure, light, and voltage. Measurement of the variability of any parameter should be typical of that encountered during a time frame of intended application. Consecutive measurements can exhibit only the extreme minimal effect of consecutive variability; they cannot demonstrate the transformation of the systematic errors of the different environments of different days into the desired estimate of the random error of haphazard choice of the particular influences affecting the blank at any future time. Ten blank measurements on 10 different days would be far more realistic. A control chart of blanks would provide the required data. Analysts must give up the notion that minimum variability indicates superior work; typically it will indicate unrealistic foresight and self-censoring.

The insidious effect of internal "improvements" in analytical methods should not be overlooked. All analysts can "improve" methods by shortening times and eliminating steps. Such changes are often taken with a minimum of comparison of the old and new version and justified on the ground that "it cannot possibly affect the results." Very often the effect cannot be seen in a series of comparisons conducted simultaneously. It is only noted that a drift has occurred after a period of time. What happens can be referred to as the picket fence effect -- where one picket is used as the pattern for the next. Adjacent pickets appear to be identical, but the first and last ones are markedly different! Laboratory management is deficient if it permits analysts to change the method, no matter how minor the change may appear, without a good reason, such as accommodating a new matrix. Unapproved changes are contrary to good analytical practices and to good quality control procedures.

Reference Materials

Reference materials are critical for evaluating the reliability of analytical results, for testing the proficiency of analysts and laboratories, and for demonstrating and maintaining adequate performance. For quality control there are only two essential requirements for a reference material: (1) homogeneity, and (2) stability. Although it is desirable that a reference value for the analyte(s) be available, this is not a critical requirement for routine quality control.

Homogeneity is achieved by adequate reduction in size and fineness and by thorough mixing. These requirements are usually achievable with one of the many types of mechanical equipment now available, sometimes combined with operation at very low temperature. Storage at low temperatures and avoidance of light, air, and moisture will usually solve most stability problems, even estimation of microbial populations.[13]

Plotting results from analysts or laboratories with time and applying calculated or historical warning limits of 2 standard deviations and action limits of 3 standard deviations provide the information required for internal quality control.

When results are to be compared over a long period of time, encompassing several reference materials, or to be compared with those from other laboratories, a reference value should be supplied, even if it is an arbitrary value, for plotting. From the plotted values, a standard deviation can be calculated which can be used as the basis for any desired quality control interpretation. There are 4 possible candidates: (1) the standard deviation itself; (2) the z-score, which is merely the ratio of the difference any given value has from an assigned mean (reference, arbitrary, or consensus) to the reference standard deviation; (3) the ISO repeatability limit (r), which has to do with the replicability of duplicates; and (4) the HORRAT score, which relates a variability to historical values. Each has its own advantages and disadvantages, depending on how it will be used.

Standard Deviation

This may be dismissed immediately as a basis for

interpretation of quality control parameters;   it is essential for calculating other more useful parameters. The standard deviation is a "floating" value because it applies only to the data it is associated with. A chemist must immediately relate it to a mean in order to place it in proper perspective as a measure of variability.

The z-Score

This index of variability answers the question "How many standard deviation units is a particular value from its associated mean?" Several other quantities must also be identified simultaneously: (1) the value of interest -- which may be an individual analytical result, a mean of a number of results, a mean from a laboratory, a mean for a method, etc., (2) the reference value -- where you are measuring from -- which may be a reference value, a mean of means, a consensus value, a "best estimate," etc., and (3) the reference standard deviation, which is the basic unit of measurement.

This is the current procedure advocated in the IUPAC harmonized protocol for proficiency testing of chemical laboratories, based upon recent experience in the United Kingdom.   Unless the reference values and standard deviations are preassigned, there is considerable room for arbitrary assignment of target values.   However, the results are easily interpreted as a normalized control chart, i.e., they are easily read in terms of z: a value >3 is unacceptable and a value between 2 and 3 is borderline.

The ISO Repeatability Limit, r

This value, which is 2.8 times the repeatability standard   deviation,   indicates   how   closely   two determinations conducted simultaneously should check each other 19 times out of 20.   Statistically, it is known as the maximum tolerable difference.   It is very much an internal quality control indicator and its usefulness depends   directly   on   the   stability   of   the   standard deviation. Because results from simultaneously conducted determinations should be close together, actual minor differences can be magnified greatly.   The great disadvantage of this parameter is that the two results on which it is based are not independent and therefore have little   relationship   to   what   would   be   expected   for independent results obtained on different days.   The corresponding reproducibility limit, R, applying to among-laboratories variability, does not have this disadvantage. The   new   ISO   5725   is   introducing   an   intermediate variability parameter to handle this problem.

The HORRAT Value

This is the ratio of the RSD (found) to the RSD predicted from the Horwitz equation.   It is of greatest value for among-laboratories variability as a starting point when no other information is available for an analytical system. In developing this curve we also tracked the within-laboratory variability, RSD_r, but did not pay much attention to it because it was so variable. But some information is better than no information if you are dealing with a new subject; consequently, 0.5 times the values calculated by the curve is a good starting point for within-laboratory precision. We always add the caveat "in the absence of overriding information." This, of course, raises the interesting question of what is overriding information. In looking at this point, we immediately saw that when you are dealing with high variabilities, of the order of RSD_r = 10-50%, there is rarely enough information to become overriding in the sense of statistical significance. We now realize that it takes an enormous number of data sets (possibly of the order of several hundred) to provide sufficient statistical "power" to distinguish an actual calculated estimate of RSD_r from any postulated target RSD_r. Therefore, any small data set of 3-6 RSD_r estimates, as provided by a single method-performance study, would be well within the prediction limits we postulate from our curve. The cautions regarding use of the model really apply to the estimates from single studies used as prediction limits and not to the estimate equation derived from the historical, pooled data. Not until an internal quality control program over a period of time has provided sufficient data to establish stable control limits should the equation-derived values be superseded. Otherwise overly optimistic limits may be established.

5   CONCLUSION

We suggest some minor, substantive revisions in the harmonized IUPAC-1987 protocol as applied to method-performance interlaboratory studies on the basis of five years of experience with its application in practical work: Remove the double split-level design, amplify the definition of material, and apply a slightly more liberal outlier removal treatment.

With respect to a harmonized protocol for internal quality control systems, we suggest acceptance of the proposed "z-score" based upon several years of experience in the United Kingdom. However, the assignment of the target standard deviation used in the denominator of the parameter should be the subject of scrutiny during the next few years to determine the "best" source for this value. We suggest a denominator, target standard deviation based upon one-half the value calculated from the Horwitz curve.

REFERENCES

1.   W. Horwitz, Pure Appl. Chem., 1988, 60, 855.

2.   W. Horwitz, J. AOAC Int., 1992, 75, 368.

3.   W. Horwitz, Pure Appl. Chem., 1990, 62, 1193.

4.   W. Horwitz, W. and R. Albert, "Biologically Related
     National Institute of Standards and Technology
     Standard Reference Materials: Variability in
     Concentration Estimates," ACS Symposium Series 445,
     K.S. Subramanian, G.V. Iyengar, and K. Okamoto, Eds.,
     American Chemical Society, Washington, DC, 1991, pp.
     50-73.

5.   E.S. Gladney, B.T. O'Malley, I. Roelandts, and T.E.
     Gills, Standard Reference Materials: Compilation of
     Elemental Concentration Data for NBS Clinical,
     Biological, Geological, and Environmental Standard
     Reference Materials, National Bureau of Standards
     Special Publication 260-111, Superintendent of
     Documents, Washington, DC 20402-9325, 1987.

6.   P. Kelly, J. Assoc. Off. Anal. Chem., 1990, 62, 855.

7.   W. Horwitz, L.R. Kamps, and K.W. Boyer, J. Assoc.
     Off. Anal. Chem., 1980, 63, 1344.

8.   W. Horwitz and R. Albert, J. Assoc. Off. Anal. Chem.,
     1991, 74, 718.

9.   Analytical Methods Committee, Analyst, 1992, 117, 97.

10. R.J. Mesley, W.D. Pocklington, and R.F. Walker,
     Analyst, 1991, 116, 975.

11. W. Horwitz, Anal. Chem., 1982, 54, 67A.

12. W. Horwitz, R. Albert, M.J. Deutsch, and J.N.
     Thompson, J. Assoc. Off. Anal. Chem., 1990, 73, 661.

13. M. Peterz and P. Norberg, J. Assoc. Off. Anal. Chem.,
     1983, 66, 1510.

Measures of Internal Quality Assurance to Be Taken at the Occurrence of Extreme Results

Ch.-G. de Boroviczény

INSTITUT FÜR STANDARDISIERUNG UND DOKUMENTATION IM
MEDIZINISCHEN LABORATORIUM (INSTAND) E.V., HASLACHERSTR. 51,
D-7800 FREIBURG, GERMANY

1 INTRODUCTION

Unexpected but reliable and exact extreme results in medical laboratories are of special interest for the clinician, as they may give him new aspects in regard of his patients. To ensure the required exactness measures of quality assurance must be taken, as otherwise the unavoidable scatter is undefined, and undetected systematic errors and/or gross faults may occur.

2 DEFINITIONS

Extreme results are defined statistically as well as clinically. The two aspects are combined, and the resulting narrowest limits taken.

Statistical Definition[1]

Statistically extreme results are results within the lowest and highest percentile of all results observed in a defined laboratory. That means, that at least 1000 subsequent results of different patients are collected in the own laboratory, and the 10 lowest and 10 highest results marked. The remaining results are the normal, borderline and pathologic, but *statistically not extreme* results. Extreme results are outside the range of the originally 11th lowest and 11th highest result, i.e. the first and the last percentile[2]. This range may be evaluated when results are listed in order of their numerical value by a computer, but it is also practicable, to mark and count enough very low and very high results manually in the lab-journal.

Clinical Definition[3]

Alarming results are clinically defined extreme results. They have to be reported to the clinician immediately, as they are indicating a possible severe danger for the patient. The borders of alarming numerical results are listed in table 1. Non-numerical extreme results are e.g. positive cross-matches, acetonuria.

Table 1. Alarming numerical results.[4]

Analyte	Lower alarming	Upper borderline	Unit
P/S-chloride	80	125	mmol/l
P/S-potassium	3	6	mmol/l
P/S-sodium	130	160	mmol/l
P/S-creatinine	–	3	mg/dl
B/P/S-glucose	40	300	mg/dl
P/S-urea	–	120	mg/dl
P/S-amylase (maltetetraose)	–	50	U/l
P/S-creatinkinase (CK)	–	100	U/l
B-haemoglobin	6	–	g/l
B-haematocrit	0,25		l/l
B-red bloodcell count (RBC)	2		TE/l
B-white bloodcell count (WBC)	1,5	50	GE/l
B-platelet count	50		GE/l
P-Quick during cumarin therapy	10	30	%
P-Quick without premedication	30	–	%
P-partial thromboplastine time	–	100	s
P-fibrinogen	100	–	mg/dl

3 MEASURES

Whenever laboratory findings are outside the borders of extreme or alarming results, special quality assurance measures have to be taken.

Reduplicate Analyses[5,6]

Reduplicate analyses should be performed with the patient sample and with a control sample using repeatability and reproducibility conditions.

Repeatability means a reduplicate analysis within *short-est* possible time, performed by the *same* person, using the *same* method, instrument, calibrator and reagent. The standard deviation of many analyses by repeatability conditions indicates the minimum deviation achievable in that particular laboratory.

Reduplicate analyses using conditions of repeatability are performed by the technician, as soon as the extreme result is noticed.

Reproducibility conditions are given, if as many parameters as possible are varied: the second analysis should be performed by *another* investigator, using *another* method, instrument, calibrator and reagent.

In some cases, e.g. during night time, only an incomplete reproducibility is practicable, but at least a different instrument and method should be used, i.e. the routine versus the stat working place.

Table 2. Conditions of Repeatability and Reproducibility[7]

Repeatability	Reproducibility
Same method	Different methods
Same instrument	Different instruments
Same calibration	Different calibrations
Same reagents	Different reagents
Same technician	Different technicians
Shortest possible interval	Remarkable time interval

The most efficient reproducibility control is the sample exchange with a laboratory in the neighbourhood. This should be performed routinely several times a week, using samples taken at random as well as all samples with extreme results. If significant differences should be observed between the results of the two laboratories, *each* should re-examinate their own methodology.

The comparison of the actual results of reduplicate analyses have to be made with the appropriate statistics of the laboratory[8]. By communication with the clinician they should be compared to the clinical observations.

Other Measures

All further measures of internal and external quality assurance, as listed in table 3, should be used by the laboratory. The *participation in surveys* and regular *statistical evaluation* of the obtained control results are stressed, and the *complete documentation* of all measures taken.

Highly Accurate (Especially Assured) Results[9,10,11]

In some cases, e.g. if results are doubted by the clinician, or to set target values for home-made control samples, a more sophisticated method is used: two technicians are performing reduplicate analyses using conditions of repeatability on two days and working places, i.e. different methods, instruments, calibrators and reagents.

An analysis of variance is performed with the 8 single results[12]. If the zero-hypothesis has not to be rejected, the all-over mean should be accepted as an highly accurate, especially assured result.

The laboratory must be able to answer the question, if *differences* between results of yesterday and today are *significant*[13] or not? This question is of special interest in connection with extreme/alarming results. To be able to answer this question, a *comprehensive, systema-*

Figure 1. Working scheme:

```
1. result\ 1.technician, 1.place\
2. result/                       |
                                 | 1.day\
3. result\ 2.technician, 2.place/       |
4. result/                              |
                                        | especially
                                        | assured result
5. result\ 2.technician, 1.place\       |
6. result/                       |      |
                                 | 2.day/
7. result\ 1.technician, 2.place/
8. result/
```

Table 3. Measures of quality assurance[14]
```
------------------------------------------------------------
```
| Internal measures | External measures |
```
------------------------------------------------------------
```

Preventive measures:

Internal measures	External measures
Selection of staff	Education
Regular int. conferences	Postgraduate training
Selection of methods	Law and standards
Selection of instruments	Instrument maintenance by
Working prescriptions	the manufacturer
Instrument manual	
Int. instrument maintenance	
Instrument protocols	
Calibration curves	
Control of reagents	
Definition of ranges, borders	
Identity control	

Control measures:

Internal measures	External measures
Multiple determinations	Sample exchange with neigh-
- repeatability condition	bouring laboratories
- reproducibility condit.	Participation in surveys
- plus patient samples by	Inspection by independent
random for short series	experts
- stability (jogging run)[15]	
- analysis of control samples	
- use of control charts	
Plausibility of single results	
- delta check	
- control of extreme results	
- connected results (e.g.GOT/GPT)	
- connection to clinical status	
Plausibility of series	
- repetitions of samples by random	
- Hoffmann-Vaid (truncated series)[16]	
- median values	
- borderline quotients	

Consequence measures:

Internal measures	External measures
Internal discussions	Certification by EQA
Internal instrument repair	Instrument repair by manu-
Updating internal statistics	facturer
Revision of int. instruction	
Replacement of obsolete methods	

tic and complete quality assurance program has to be performed day after day for all methods used in the laboratory for routine or stat analyses, furthermore with all reserve instruments, *and the results of controls have to be evaluated statistically, updating the statistics regularly.*

References

1. K-G.v. Boroviczény, R. Merten, U.P. Merten (ed), "Qualitätssicherung im Medizinischen Laboratorium", INSTAND Schriftenreihe 5. Springer-Verlag, Berlin, Heidelberg, New York, London, Paris, Tokyo, 1987, ISBN 3 540 13496 4. Chapter 2.6, p. 141.

2. A. Naiman, R. Rosenfeld, G. Zirkel, "Understanding Statistics" McGraw-Hill, New York, St.Louis, San Francisco, Auckland, Bogotá, Düsseldorf, Johannesburg, London, Madrid, Mexico, Montreal, New Delhi, Panama, Paris, São Paulo, Singapore, Sydney, Tokyo, Toronto, 2nd.ed., 1977.

3. K-G.v. Boroviczény, R. Merten, U.P. Merten (ed), "Qualitätssicherung im Medizinischen Laboratorium", INSTAND Schriftenreihe 5. Springer-Verlag, Berlin, Heidelberg, New York, London, Paris, Tokyo, 1987, ISBN 3 540 13496 4. Chapter 2.7b, p. 176.

4. K-G.v. Boroviczény, R. Merten, U.P. Merten (ed), "Qualitätssicherung im Medizinischen Laboratorium", INSTAND Schriftenreihe 5. Springer-Verlag, Berlin, Heidelberg, New York, London, Paris, Tokyo, 1987, ISBN 3 540 13496 4, Chapter 2.6, p.136.

5. S.M. Lewis, R.L. Verwilghen, "Quality Assessment in Haematology", Ballière Tindall, London, Washington, Toronto, North Ryde, Tokyo, 1988, ISBN 0 7020 1322 6, pp. 43-73.

6. K-G.v. Boroviczény, R. Merten, U.P. Merten (ed), "Qualitätssicherung im Medizinischen Laboratorium", INSTAND Schriftenreihe 5. Springer-Verlag, Berlin, Heidelberg, New York, London, Paris, Tokyo, 1987, ISBN 3 540 13496 4, Chapter 2.4, p. 95.

7. K-G.v. Boroviczény, R. Merten, U.P. Merten (ed), "Qualitätssicherung im Medizinischen Laboratorium". INSTAND Schriftenreihe 5. Springer-Verlag, Berlin, Heidelberg, New York, London, Paris, Tokyo, 1987, ISBN 3 540 13496 4, Chapter 7.1, p. 906.

8. K-G.v. Boroviczény, R. Merten, U.P. Merten (ed), INSTAND Schriftenreihe 5. Springer-Verlag, Berlin, Heidelberg, New York, London, Paris, Tokyo, 1987, ISBN 3 540 13496 4, Chapter 2.4, p. 96f.

9.   K.G.von Boroviczény, <u>Exc.Med.Int.Congr.Series</u>,
     1975, <u>384</u>, 239-241.

10. The Sci.Org.Comm.of The 5th Internat.Symp. on Qua-
     lity Control, "Proceedings", Simul Internat.Inc.,
     Japan, 1984, p. 208.

11. R. Merten et al. (ed), "Zielwert, Sollwert, Ziel-
     bereiche in der Laboratoriumsmedizin", INSTAND
     Schriftenreihe 3, Springer-Verlag, Berlin, Heidel-
     berg, New York, Tokyo, 1984, ISBN 3 540 13455 7.
     Chapter 5.1, p. 130.

12. R. Merten et al. (ed), "Zielwert, Sollwert, Ziel-
     bereiche in der Laboratoriumsmedizin", INSTAND
     Schriftenreihe 3, Springer-Verlag, Berlin, Heidel-
     berg, New York, Tokyo, 1984, ISBN 3 540 13455 7,
     Chapter 5.1, p.131.

13. S.M. Lewis, R.L. Verwilghen, "Quality Assessment in
     Haematology", Baillière Tindall, London, Washing-
     ton, Toronto, North Ryde, Tokyo, 1988, ISBN 0 7020
     1322 6, pp. 43-73.

14. K-G.v. Boroviczény, R. Merten, U.P. Merten (ed),
     "Qualitätssicherung im Medizinischen Laboratorium",
     INSTAND Schriftenreihe 5, Springer-Verlag, Berlin,
     Heidelberg, New York, London, Paris, Tokyo, 1987,
     ISBN 3 540 13496 4, Chapter 2.1, p. 44.

15. The Sci.Org.Comm.of The 5th Internat.Symp. on Qua-
     lity Control, "Proceedings", Simul Internat.Inc.,
     Japan, p. 203.

16. R.G. Hoffmann, "Establishing Quality Control and
     Normal Ranges in the Clinical Laboratory", Exposi-
     tion Press, New York, 1971, ISBN 0 682 47167 4.

New Approaches to Automation and Standardisation of Tests for Platelet Storage Lesion and Activation State

M.J. Seghatchian and J.F.A. Stivala
QUALITY DEPARTMENT, NORTH LONDON BLOOD TRANSFUSION CENTRE (NLBTC), COLINDALE AVENUE, LONDON NW9 5BG, UK

INTRODUCTION

Blood Transfusion Services, like any pharmaceutical manufacturing, are striving towards attaining the highest quality standards by compliance with the principles of BS5750 or its equivalent ISO9000 series, hence ensuring that a formalised quality system is operational in the organisation. At NLBTC our system focuses not only on operational procedures, results, products, services, training and continuing education, but also on less apparent aspects of quality concepts such as quality as a way of life reflected in the mission statement, and continual striving for quality improvement, in the face of economic restraints.

In pharmaceutical manufacturing, compliance with current GMP requires systems to be in place that ensure that no defective product reaches the market. Real quality standard is effectively achieved when it is undertaken with the objective of satisfying the customer's need as the most important criterion. A similar concept is applicable in the transfusion setting where extensive efforts are directed towards the reliability in both product and services with two of their essential attributes, accuracy of information and timeliness.

The difficulty which one faces with individual donations is that each unit of blood has to be considered as a discrete batch and due to population variability in donors, even 100% testing of products would not assure that the real standard of quality is always achieved reliably in terms of satisfactory outcome for prospective patients. Moreover, clinical trials are not only inconvenient to the patient but also inappropriate if a standard testing procedure reflecting the state of the art technique is not employed.

Too often a blood component, collection/processing equipment or a process once well adjusted to meet a certain requirement fails to provide the same performance after a short time while users are all too frequently unaware of it, in particular if infrequent testing is applied to predict such pitfalls. The current production, storage and testing of platelet concentrates (PC) fulfil this scenario as not only processing introduces further heterogeneity in the source blood but the current platelet function tests are largely time consuming, lack sensitivity and are poorly validated against standards, where these exist.

We describe a set of laboratory tests which assist in the assessment of platelet heterogeneity and performance characteristics of various equipment or processes used, by lending themselves to automation, assist in large scale pre-screening of PC, helping in the standardisation and harmonisation process. Most of these developments arose from the innovative work carried out by staff in our Quality Department in collaboration with other transfusion centres. Obviously we would like to see widespread acceptance and implementation of these new approaches/developments, preferably under the auspices an international organisation such as ISO with its universally recognised expertise in uniformity and harmonisation of clinical laboratory testing. The following areas are briefly described in this paper.

The Need for Automation in Quality Assurance of Platelet Concentrates

Quality Assurance (QA) is one of the most important and least appreciated activities in blood transfusion and clinical laboratories. The reason for this shortcoming possibly rests on the rather dull and laborious function testing, calculation, documentation and record-keeping that it involves. Nevertheless, the pressure is steadily increasing on laboratories to effectively and rapidly manage the two basic functions of a QA programme: firstly to evaluate QA data as it is generated and secondly to produce data summaries promptly for product release and peer comparison.

In blood transfusion settings, while impressive strides have been made towards harmonisation of mandatory testing for donor screening and the operational aspects of blood component collection, processing and storage, there has been comparatively little progress towards standardisation/automation of source, intermediate and final product testing. With the advent of automated technology, e.g. automated cell counters, microplate ELISA, it is now possible to introduce reliable large scale screening tests to differentiate viable (active) from less viable and/or reject products based on function testing.

Apart from reliability of these well established new generation testing systems which by derivation could lead to the establishment of a standardised procedure, tests for platelet function should embrace the following attributes:

- Small sample volume to minimise product loss (non-destructive testing).

- Simple and rapid for large-scale pre-release testing.

- Relevant and accurate reflecting one of the important functional properties.

- Easy to perform lending itself to automation and standardisation.

An Automated Analyser for Assessment of Platelet Storage Lesion (PSL)

The three desirable requirements for an automated cell analyser for PSL are:

- An appropriate reagent system that maintains cells as close as possible to their near native state.

- A measurement system capable of direct and simultaneous evaluation and differentiation of the cellular content without need for additional dilution or special stains so that the native volume, cytoplasmic content and surface characteristics of cells remain relatively intact.

- A comprehensive data management facility to provide precise mapping of multicellular characteristics with displays that allow operators discrete analysis of cellular population with ease.

New automated analysers such as the Technicon® H*1 and H*2 are designed to determine such indices as platelet count, mean platelet volume (MPV), platelet distribution width (PDW, reflecting the coefficient of variation of the platelet volume histogram) and plateletcrit computed in a manner similar to that used for red cell haematocrit. These indices are accurately measured by laser scattered light at two angles (2.5°-3.5°, 5°-15°) enabling direct assessment of cellular morphology, based on shape changes and cytoplasmic content. The accuracy of the data generated is enhanced by hydrodynamic focusing of cells in single file in a flow cell before analysis. The analyser manages the data efficiently within 1 minute with less than 1% carryover and wide range of linearity, eliminating the need for pre-dilution.

The leucocytes (WBC) are differentiated in such a manner that disparity between basophil lobolarity and peroxidase activity provides an excellent indicator for the presence of platelet clumps. Such information is relevant in assessing the quality of platelet concentrates where WBC levels are required to be minimised, with a high level of confidence in the measurement by automated system.[1,2]

We have observed that automated cell counters, though designed for use with EDTA anticoagulated blood in clinical laboratories, can be used for assessment of changes in cellular indices of citrated blood samples, though one must ensure appropriate interpretation of the test result generated and their usage in patient care. At NLBTC, the use of a Technicon® H*1 during the past few years has provided a wealth of information on circulating platelet aggregates, spontaneous aggregation, loss of spontaneous aggregation within 48 hours, discoid/spheroid conversion during storage, microvesiculation/fragmentation due to a very high or low pH (>7.6 or <6.6).

Other influencing variables in quality include 37°C to 4°C temperature recycling induced reversible/irreversible morphological changes in platelet size, vigorous agitation, hyper or hypotonic shock and mechanical trauma induced cellular injury, and centrifugation/filtration induced changes in the distribution of different populations of platelets in red cell, PC and plasma components. These issues are directly relevant to quality and must be built into the design from the start through processing, storage and shipment, even before manufacture of the product. One way of achieving it is the implementation of stringent in-process QC followed by trending on the basis of random testing, as a means of raising staff awareness of essential (minimum specification) and desirable (idealistic specification) requirements.

EDTA Reveals the Secrets of Platelet Morphological/Functional Integrity

The measurement of platelet indices has only recently achieved respectability in the clinical field. We have observed that the addition of 0.5 mL citrated PC to dipotassium EDTA sample tubes, followed by the assessment of cellular indices of paired samples (± EDTA) provide an extremely useful test of platelet function on the basis of the difference between MPV (dMPV).[3,4] In addition, accurate and quantitative information on the activation/aggregation state of fresh sample is obtainable by the use of the dPLT, the difference between PLT of paired samples.

EDTA improves the validity of leucocyte/erythrocyte content of sample as clumps/aggregates in citrated samples often misclassify in automated cell counters; these are dispersed in EDTA containing samples. EDTA also reveals other secrets of platelet integrity, for example, aged and spheroid platelets do not respond to EDTA as measured by dMPV and dPLT, platelets which have undergone shape changes, (i.e. cold shock, thrombin activated) also fail to swell when added to EDTA. The influence of these variables which are known to have a dramatic effect on the quality during collection and processing stages, collectively called platelet storage lesion, are easily measured by automated cell counting.[1-4]

Recently[3], we have validated dMPV with other conventional tests for platelet function; the high correlation ($r > 0.90$) suggests that dMPV change due to EDTA or storage is of essential value in the standardisation of quality and should be included as an essential part of the specification in both laboratory and clinical fields of platelet abnormality. (See Table 1)

Automation and Data Management of Tests based on Aggregometry Principles and Release Reactions

Platelet aggregation studies are carried out in clinical laboratories for diagnostic purposes and in blood transfusion for acquiring a better understanding of the most effective ways of collecting, processing and storing of PC to minimise platelet activation/release reaction and clumping. Conventional light transmission instruments based on the Born Principle are used; the data are recorded in the form of superimposed traces for visual comparison of time taken for maximum light transmission (T_{max}) and the rate of primary and secondary

Table 1. Relationship between conventional function tests and dMPV assessed by two types of cell counters. (Adapted from Vickers, *et al.*[3])

dMPV as measured by	Technicon® H*1[a]		Sysmex® M2000[b]	
	r	p	r	p
20μM ADP *vs.* dMPV	0.81	0.0436	-0.91	0.0091
10μM ADP + 100μg/mL collagen *vs.* dMPV	0.81	0.0414	-0.90	0.0104
pH *vs.* dMPV				
Days 0 - 7	0.76	0.0670	-0.71	0.1070
Days 3 - 7	1.00	0.0002	-0.86	0.1899
HSR *vs.* dMPV	0.85	0.0250	-0.77	0.0647
vWF:Ag *vs.* dMPV	-0.92	0.0070	0.73	0.0061
βTG *vs.* dMPV	-0.96	0.0013	0.96	0.0018
LDH *vs.* dMPV	-0.95	0.0029	0.87	0.0186
dMPV *vs.* age	-0.98	0.0002	0.97	0.0007
20μM ADP(%) *vs.* age	-0.78	0.0600	-0.83	0.0310
10μM ADP + 100μg/mL collagen *vs.* age	-0.80	0.0488	-0.95	0.0029
pH *vs.* age				
Days 0 - 7	-0.76	0.0670	-0.83	0.0320
Days 3 - 7	-0.94	0.0760	-0.99	0.0180
HSR *vs.* age	-0.91	0.0080	-0.98	0.0005
vWF:Ag *vs.* age	0.97	0.0080	0.96	0.0013
βTG *vs.* age	0.98	0.0004	0.98	0.0003
LDH *vs.* age	0.95	0.0027	0.94	0.0038

[a]　　Determined at the North London Blood Transfusion Centre, Colindale
[b]　　Determined at the South Western Regional Blood Transfusion Centre, Bristol

aggregation (slope), or by calculating the rates. The visual comparison of data is qualitative at best and calculation of aggregation parameters are time consuming. Moreover keeping track of original data in trace form and its retrieval is highly cumbersome.

At NLBTC we have overcome some of these problems using a dual channel Chrono-Log® Lumi-Aggregometer based on impedance principles. The instrument is equipped with an integrator which significantly increases the efficiency and ease in performing the experiment, eliminating the tedium involved of calculating aggregation parameters and rapidly provide data in tabular and graphic forms, as well as being a more effective method of data storage, retrieval and comparison. The results of aggregometry convey two sets of new information, firstly a biphasic nature of decay in aggregation during storage showing a rapid drop within 24 hours followed by slower rate of response, and secondly, the development of an inhibitor of aggregation in plasma during storage as platelets are rejuvenated upon resuspension in fresh plasma. This is supported by the release of ATP/ADP measured by luminescence. The capacity of releasing ATP/ADP upon stimulation is one of the essential requirements of platelet physiological function and should be retained until the end of the shelf life, requiring large scale screening.

Towards Automation of Platelet Function Testing with New Generations of Molecular Markers of PSL

Microplate technologies are now well established procedures in blood transfusion services with proven reliability for timely mandatory screening tests. The assessment of low grade proteolytic activity generated due to contact with artificial surfaces or tissue factor release, and kallikrein or thrombin generation during collection, processing and storage is also facilitated with microplate systems, using chromogenic substrate analogues of serine proteases.

Recently[1,5] new ELISA assays for both von Willebrand Factor collagen binding activity (vWF:CBA), vWF:Ag and glycocalicin (a major segment of the glycoprotein 1b) have become available and applied to quality monitoring of the PC. For adequate functional support, the recovered platelets must remain in their native form during processing and up to five days storage and preserve their ability to adhere to the site of injury, undergo release reactions, and to correct primary haemostasis and arrest bleeding. Both vWF and GP1b are essential components of haemostatic modality, in fact vWF is known as an adhesive protein essential in bridging platelets through GP1b to exposed sub-endothelium surfaces.[1]

The structure/function relationships of platelet membranes, in particular the exposure of GP1b and vWF to platelet surface upon activation, are determined by flow cytometric principles. There has been some concern whether the use of flow cytometry provides accurate information on PSL as variable degrees of membranous exocytosis and endocytosis occurs in various stages of processing.[1] Other strategies for the estimation of platelet activation/storage lesion are the assessment of released/cleaved glycocalicin and intracellular released/releasable vWF in supernatant plasma of PC during storage. The latter molecular marker can be evaluated both by functional assay (vWF:CBA) and immunologically (vWF:Ag) by ELISA providing an unique test for assessing vWF activity states based on the ratio of vWF:CBA *vs* vWF:Ag. The application of these procedures to routine and apheresis PC stored up to eight days revealed that the presence of even low levels of WBC in PC dramatically accelerates the proteolytic fragmentation of vWF identifiable by non-parallelism in biometric assay of both vWF:CBA and vWF:Ag as well as changing in their ratio during storage. This is supported independently by dMPV measurements showing in some cases a relatively small or negative value within the shelf life (day 4-5 storage).

The release of cellular enzymes such as calpain, elastase and chymotrypsin like enzymes are also associated with elevated levels of both vWF:Ag and glycocalicin in microvesicular and soluble form. In most conditions associated with PSL, the level of these molecular markers correlate well with each other and dMPV during storage.[5,6] Hence new sets of both morphological and functional markers for PSL are now available facilitating automation standardisation of platelet storage lesion, helping in large scale screening.

FUTURE TRENDS

Quality issues in blood transfusion medicine will be of continued interest to donors and patients as well as regulatory agencies. Reliable tests for platelet quality should have at least two essential attributes, accuracy and timeliness, to prospective patients who expect the safest and most efficacious product and clinicians who use the laboratory data for decision making, therefore expecting the reported information to be reliable and relevant. Taking into account the biological variation in the blood donor population and keeping in mind that a defective product can cause damage to a patient; failure to conform to minimum accepted standards of practice would place organisations in an unfortunate position. Accordingly, one might expect that in the near future the need for a real platelet quality specification, covering both morphological changes and functional integrity, will attract the attention of the regulatory agencies. This issue is currently of particular relevance due to legislative changes affecting transfusion services in the UK and the European Community. These new generation tests, which lend themselves to automation, are here to stay and their universal application will not only help in harmonisation to a certain product norm based on relevant clinical function which satisfy the outcome of the platelet transfusion, but also eliminate existing controversy based on small clinical trials. Technology in this field is rapidly changing; the challenge is to utilise these changes for the good of transfusion medicine.

REFERENCES

1.   M J Seghatchian, Blood Coagulation and Fibrinolysis, 1991, 2, 357.

2.   M J Seghatchian and J F A Stivala, 'Quality Assurance in Transfusion Medicine', CRC Press Inc, Boca Raton (FL), USA, 1992, Volume 1, Chapter 1, p. 1.

3.   M V Vickers, A H L Ip, M Cutts, N P Tandy and M J Seghatchian, Blood Coagulation and Fibrinolysis, 1991, 2, 361.

4.   M J Seghatchian and B Brozovic, Blood Coagulation and Fibrinolysis, 1992, 3, 617.

5.   H Bessos, W G Murphy and M J Seghatchian, Blood Coagulation and Fibrinolysis, 1991, 2, 373.

6.   H Bessos, M J Seghatchian, M Cutts and W G Murphy, Blood Coagulation and Fibrinolysis, 1992, 3, 633.

Reference Materials, Reference Values of Statistical Tests in the Quality Assurance Schemes of Analytical Laboratories

L. Paksy[1] and M. Parkany[2]

[1]METALCONTROL KFT MISKOLC, PO BOX 557, H-3510, HUNGARY
[2]ISO CENTRAL SECRETARIAT, GENEVA 20, CASE POSTALE 56, CH-1211, SWITZERLAND

1 INTRODUCTION

The use of statistical tests is indispensable in the quality assurance schemes of analytical laboratories, as well as in analytical methodology as a whole[1]. Without these tests, decisions regarding the validity of analytical results, the adoption of a method as a standardized one, or the monitoring of the performance of a laboratory etc., cannot be made.

However, all recommended statistical tests (see for example in reference [2]) such as t-tests, F-tests, χ^2-tests, etc., are based on comparisons with known or commonly accepted values.

A measurement can be characterized by values of $\bar{x}$, s and n, where $\bar{x}$ is the estimate of the mean value, s the standard deviation, and n the number of parallel determinations. Any control of a measurement requires the values of $\bar{x}$ and s as well as the acceptance of these values.

According to ISO "Harmonized Proficiency Testing Protocol"[2], $\bar{x}$ can be tested by means of certified reference materials, whereas for s four possibilities are mentioned: arbitrary fixation, prescription, reference to a validated methodology, or to a generalized model such as the Horwitz curve[3].

According to ISO Guide 30[4], a certified reference material is a material, "...which is accompanied by a certificate, one or more of whose property values are certified by a procedure which establishes its traceability to an accurate realization of the unit in which the property values are expressed, and for which each certified value is accompanied by an uncertainty at a stated level of confidence".

Certified reference materials (CRMs) for chemical analysis are characterized by concentration values, and they can be used therefore for the control of the mean value $\bar{x}$ of an analytical process or an instrument calibration. A common requirement is the homogeneity of these materials, and their physical and chemical characteristics should approximate those of the material to be analysed. In fact, for calibration various CRMs are needed for various matrices.

As defined in reference [2], there is a repeatability (within-laboratory) standard deviation, s_r, and a reproducibility standard deviation s_R which is also used in collaborative trials. These values are not property values of CRMs; however they are characteristic for a given analytical method used for a given analytical task (for example matrix, element, concentration).

From the point of view of both the quality assurance systems of the laboratory as well as the determination of the values of s_R, these parameters

($\bar{x}$, s_r, s_R) are necessary for proficiency testing and the acceptance of s_r is of fundamental importance. However, for its acceptance a reference value is needed and, in accordance with the above-mentioned facts, this value is likely to depend on the parameters of matrix, element and concentration, and - basically - on the analytical method.

Though generalized s_r values can also be derived from the generalized model[3] ($s_r \geq 0.5s_R$), it is well known that there are deviations from the Horwitz curve. Therefore, regarding within-laboratory standard deviations, it appears necessary to investigate the dependence of s_r on certain essential parameters: in this case, matrix, element and concentration.

We hope that clarification of this dependence will contribute to the correct determination of the reference value of s_r, or, according to ISO/REMCO N 263[2], to the correct determination of the target value of s_r. According to ISO/REMCO N 263, the target value for standard deviation is a numerical value "which has been designated as a goal for measurement quality" (ISO/REMCO N 263, §2.9).

2   REFERENCE VALUES FOR STANDARD DEVIATIONS

2.1   The importance of the use of a reference standard deviation in statistical tests

If we want to determine whether a given measured value $\bar{x}$ with a standard deviation s_x from n parallel determinations is acceptable, we use Student's t-test:

$$t = \frac{x_0 - \bar{x}}{s_x} (m)^{1/2} \tag{1}$$

It is evident that the larger the value of s_x accepted, the larger the deviation from the reference value of the concentration (i.e. from x_0) which will also be accepted. The statistics alone may be correct, indicating, however, a measurement of poor and unacceptable quality.

We need therefore a further test, and a decision on the acceptability of s_x. This the well-known F-test:

$$F = \frac{s_x^2}{s_{x,0}^2} \tag{2}$$

wher s_x is the standard deviation to be accepted, and $s_{x,0}$ is the reference value to which it is compared. (The criterion for the acceptance of s_x is the condition: $F < F_{tab.}$, where $F_{tab.}$ is the tabulated value of F for a given degree of freedom f for n parallel determinations at a probability of P%). The decision about s_x, and consequently $\bar{x}$, basically depends on the value of $s_{x,0}$.

2.2 Dependence of the relative standard deviation on matrix, element and concentration

From a practical aspect, it is useful to calculate the relative standard deviation, RSD, and to investigate its dependence on matrix, element and concentration. RSD is given by the expression:

$$\% \, RSD = \frac{s_x \cdot 100}{\overline{x}}$$

Let the subscripts r and R denote repeatability and reproducibility, respectively. Using this designation, the following investigations refer to RSD_r values.

As an example, the dependence of RSD_r values on concentration in the optical emission spectrometric analysis of steel is shown in Fig. 1.

It can be seen that:

(a) the dependence on concentration is a continuous function; the RSD values decrease with increasing concentration (see e.g. carbon and chromium);

(b) the shapes of the curves may be quite different (see e.g. curves for carbon and chromium);

(c) "irregular" curves may also occur (see e.g. molybdenum). An illustration of this behaviour is given in Fig.2. In comparison to curve **a** showing the "ideal" curve for carbon (see Fig.1), curve **b** has a different character, if it is a curve at all (further discussion is found in the next section).

Figure 3 gives more detailed information on the dependence of RSD_r on the element; irregular behaviour is shown for the elements vanadium, tungsten and nickel. These $\overline{RSD}_r$ values all refer to the same matrix (an 18/8 type steel) and the possible influence of the matrix is thus excluded.

In Fig. 4, the influence of the type of matrix is illustrated. Although they are all iron-based alloys, the basically monotonically decreasing trend (see Figs. 1,2) is disturbed by the elements vanadium, molybdenum and manganese, but not by carbon, the main alloying element of the iron alloys.

The stability of $\overline{RSD}$ values with time should also be investigated. Some examples are given in Fig. 2 (vertical lines show maximum-minimum values measured within a short time period); however, more detailed results on long-term stability are shown in Fig. 5 ("control charts"). During these long-term investigations, special attention should be given to the possibility of drift (a method is described in reference [6]).

From Fig.5, it can be seen that the variation with time does not exceed the value $2 \, RSD_{r,o}$. According to the F-test [eqn. (2)], let $s_x = 2 \, RSD_{r,o}$, $s_{x,o} = RSD_{r,o} = \overline{RSD}_r$, consequently:

$$F = \frac{2^2 \cdot RSD_{r,o}^2}{RSD_{r,o}^2} = 4 < 5.05 \; (F_{tab}; \; f_1 = f_2 = 6\text{-}1 = 5; \; P = 95\%)$$

In this case, therefore, there is no statistically significant difference between the mean value and the minimum or maximum value of RSD_r, respectively. (This does not mean a drift-free measurement).

3   CONCLUSIONS AND REMARKS CONCERNING THE USE OF
      RELATIVE REPEATABILITY STANDARD DEVIATIONS AND CRMs

Relative repeatability standard deviations - RSD_r-s - have a fundamental role in quality assurance schemes. Experimental data, as well as the corresponding ISO documents, (see reference [7]), emphasize the need for the use of "task-tailored" reference values:

"Schemes may set out to assess the competence of laboratories undertaking a specific analysis in a specified matrix (e.g. determination of lead in blood, or fat in bonemeal) rather than the general type (food analysis) mentioned"[2];

"If the repeatability standard deviation, s_r, of the standard measurement method is known, s_w can be estimated by the following procedure:
Compute the ratio

$$C = s_w^2/s_o^2 \quad \text{. . .} \quad \text{(See reference [7])}$$

Note: The value s_w refers to the intralaboratory tests, s_o is the reference value for the tests.

In case of a multielement method - such as the optical emission cited here - another question arises: are the RSD values for various elements statistically the same or different, considering the simultaneous, uniform analysis procedure?

Considering the use of a reference repeatability standard deviation - $s_{r,o}$ or $RSD_{r,o}$, respectively, the following conclusions can be drawn regarding the acceptance of a given, measured value of RSD_r:

1. The actual value of $RSD_{r,o}$ depends on the analytical method, the matrix involved, the element under investigation and its concentration.

2. The actual $RSD_{r,o}$ value should, if possible, be determined by thorough and careful measurements, using CRMs and optimum experimental conditions. An example is given in Table 1, however, in the absence of corresponding CRMs, using "in-house samples". (The outlying values in the measurement series may be due to the possible inhomogeneity of the samples.)

3. Because of the variety of analytical methods and matrices (e.g. for iron-based alloys), various CRMs are needed for various alloy types (see Fig.4), as well as elements to be analysed and concentration ranges.

4. In the absence of adequate CRMs, "in-house samples", or as a "rule of thumb" the half-values derived from the Horwitz curve [$RSD_R = 2 \exp (1-0.5 \log c)$, where c is the concentration expressed as a mass fraction] can be used.

To these conclusions, the following remarks should be added:
  - re: "dependence on concentration":
    the dependence of concentration is a continuous function, as clearly expressed in the Horwitz curve, as well as in experimental data; however,
  - re: "preferred experimental data":
    this continuous function cannot always be determined, because in a given type of alloy (or any sample to be analysed, e.g. blood, etc.) the element content varies over only a narrow concentration range.

In this connection, it must be noted that curve **b** in Fig. 2 ($RSD_{r, Mo}$ vs. $c_{Mo}\%$) can not be regarded as a real curve; rather it represents only the deviations from an "ideal" curve, like curve **a** ($RSD_{r,C}$ vs. $c_C\%$).

This means that in the spectrometric analysis of steel for molybdenum, for every type of steel the $RSD_{r,Mo,o}$ values must be determined separately.

For carbon, however, the general $RSD_{r,C}$ vs, c_C % curve can be well determined; e.g. on comparing the value of $RSD_{r,C}$ given in Table 1 with the value of $RSD_{r,C}$ in Fig. 4 at C = 1.90 % (= 0.5%), there is no significant difference for steel type K9 ($RSD_{r,C,o}$= 0.8 %; F = 0.8²₃ 0.5²=2.56<3.79 where 3.79 is the tabulated value at f_1=f_2=5 and P=95%).

For the element molybdenum of this steel type (K9), the deviation from curve **a** (see the mean value at point 4 in Fig.2) is significant.

- re: "multi-element analysis by optical emission spectrometry" it must be noted that $RSD_{r,o}$ values can differ for different elements in the same concentration range;
- re: "use of CRMs":
  To eliminate the disturbing effect of inhomogeneity, the use of CRMs is recommended (if the adequate type is available); however, the $RSD_{r,o}$ value characterizes the entire analytical procedure and not the CRMs. This requirement is more pronounced in the case of a standard method;
- re: "use of RSDs of standard methods and target values. The repeatability, i.e. the within-laboratory standard deviation, cannot be used either as the RSD_r of the standard method or as that of the target value. However, as these values are based on the measurements made in individual laboratories, the use of matrix-element method-tailored RSDs contributes basically to the improvement of the reliability and the establishment of harmonized quality assurance schemes.
- re: "use of the Horwitz curve":
  As is well known, the Horwitz curve is a generalized error function (error vs. concentration) based on many thousands of analytical results made in many hundreds of laboratories. If there is no possibility for the direct determination of the specific RSDs, the half-value derived from the Horwitz curve can be used for $RSD_{r,o}$. As in the examples given, the measured RSD_r values are always below this half-value. Regarding the optical emission spectrometric analysis of steel, this derived value can be considered as the maximum permissible error. (Evaluation: F-test).
  It must be emphasized that the use of a continuous error function is of vital importance, instead of the error functions with stepwise character applied in earlier analytical standards.

ACKNOWLEDGEMENT

The authors are most indebted to Dr. W. Horwitz and Dr. A. Head for their very useful and valuable advice and remarks.

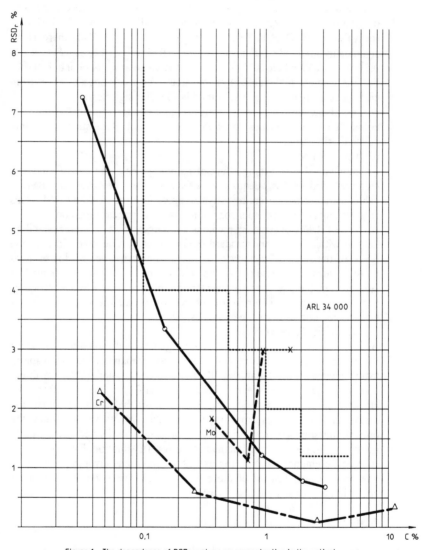

Figure 1 – The dependence of RSD$_r$ values on concentration in the optical emission spectrometric analysis of steel (using ARL 34 000 spectometer) for the elements C, Mo and Cr. Dotted line: tolerance limits according to the Hungarian standard for C-determination.

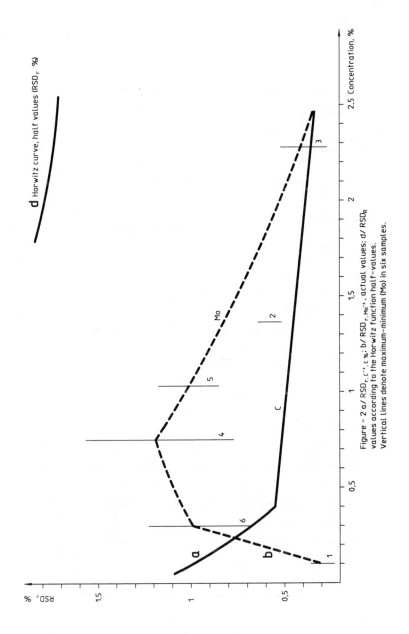

Figure – 2 a/ RSD_r, c^{-r}, c %; b/ RSD_r, Mo^{-r}, actual values; d/ RSD_R values according to the Horwitz function half-values. Vertical lines denote maximum–minimum (Mo) in six samples.

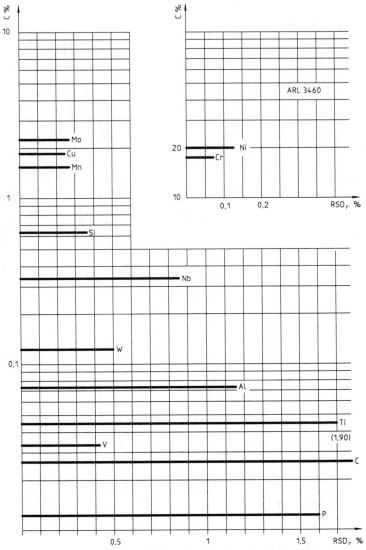

Figure 3 – The dependence of RSD_r values by various elements in a stainless
steel / ARL 3460 type optical emission spectrometer / on concentration.

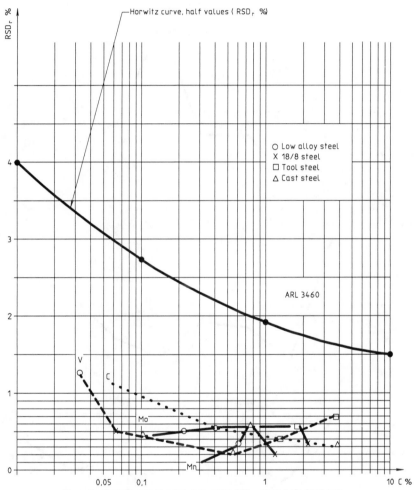

Figure 4 – The influence of matrix type on the RSD_r values / ARL 3460 type optical emission spectrometer: steel analysis /.

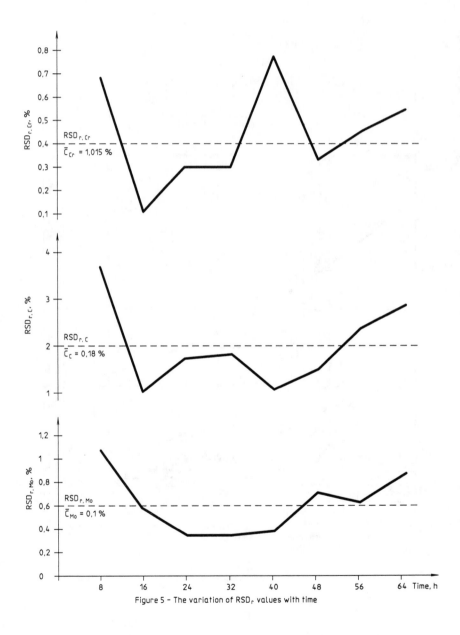

Figure 5 – The variation of RSD$_r$ values with time

Table 1.

Reference standard deviation - s_r values for C and Mo in steel K-9
(Bartlett-test for the homogeneity of the s values)

Group	n_j	s_C	s_{Mo}
1	10	0.0111	0.116
2	10	0.0170	0.0056
3	10	0.0115	0.0072
4	10	0.0252	0.0122
5	10	(0.0289)	$(0.0170)^x$
6	10	0.0179	0.0117
7	10	0.0165	0.0066
8	10	0.0088	0.0086
9	10	0.0147	0.0069
s_r		0.0152	0.0091
$s_r \%$		0.8	1.2
x_0 value/		1.90 %	0.75 % /certified
$\chi^2 /95/$		14.07	14.07
χ^2		4.50	10.82
HOMOGENEOUS ?		YES	YES

x : outlying results

REFERENCES

1.   Parkany, M, in Kivalo, P. Standardization in Analytical Chemistry Akadémiai Kiadó, Budapest, 1989, p. 64.
2.   ISO/REMCO N 263, November 1992, International Harmonized Protocol for the Proficiency Testing of Chemical Analytical Laboratories.
3.   Horwitz, W., Kamps L.R. and Boyen K.J., *Assoc.off.Anal.Chem.*, **63**, 1980, 1344.
4.   ISO Guide 30, Terms and definitions used in connection with reference materials, 1992.
5.   Horwitz W. and Albert H. ISO/REMCO N 217, Certified Reference Materials -Variability in Concentration Estimates, March, 1991.
6.   Paksy L. *Microchem, J.*, **45**, 1992, p.318.
7.   ISO/DIS 5725-4: 1992, Basic methods for the determination of the trueness and precision of a standard measurement method.

Reference Materials for Quality Control in Water Microbiology

K.A. Mooijman[1], A.H. Havelaar[2], S.H. Heisterkamp[3], and N.G.W.M. van Strijp-Lockefeer[1]

[1]FOUNDATION FOR THE ADVANCEMENT OF PUBLIC HEALTH AND ENVIRONMENTAL PROTECTION (SVM), PO BOX 457, 3720 AL BILTHOVEN, THE NETHERLANDS

[2]LABORATORY FOR WATER AND FOOD MICROBIOLOGY AND [3]CENTRE OF MATHEMATICAL METHODS, NATIONAL INSTITUTE OF PUBLIC HEALTH AND ENVIRONMENTAL PROTECTION (RIVM), PO BOX 1, 3720 BA BILTHOVEN, THE NETHERLANDS

1 INTRODUCTION

In 1986 the National Institute of Public Health and Environmental Protection (RIVM, Bilthoven, The Netherlands) started, with support of the Community Bureau of Reference (BCR, Brussels, Belgium), with the difficult task of preparing reference materials for water and food microbiology. Difficult, because of dealing with living organisms. However, in the past 6 years it was shown to be possible to prepare microbiological reference materials which fulfil the (general) requirements[1]:
a) Representative (i.e. composition and use should correspond as much as possible to the type of samples routinely examined); b) Homogeneous and c) Stable (during storage the decrease of the contamination level should be within certain limits over a certain period of time).

Although it is easier to homogenize an aqueous solution than a powder, the contents of the latter are generally more stable. Beckers et al[2] demonstrated that artificially contaminated spray-dried milk was suitable as a basic material for the preparation of reference materials for food microbiology. By mixing this 'highly contaminated' milk powder with sterile milk powder, the contamination level could be adjusted. The final reference materials consisted of gelatin capsules filled with 0.2-0.3 g mixed milk powder. This procedure was also followed for the preparation of reference materials for water microbiology. Before use, the capsule needs to be reconstituted in a certain amount of peptone saline solution. The solution represents a 'simulated water sample' which can be analysed as an ordinary water sample.

REFERENCES

1. B. Griepink, Quimica analitica, 1989, 8, 1.
2. H.J. Beckers, F.M. van Leusden, et al, J. appl. Bacteriol., 1985, 59, 35.

2 MATERIALS AND METHODS

Media

The following media were used: Lauryl Sulphate Agar
(LSA), contained Membrane Lauryl Sulphate broth (Oxoid
MM615) with 12 g agar added to 1 litre. Kenner Fecal
Streptococcus Agar (KFA, Difco 0496). 0.1 % Peptone Saline
solution (PS) contained (g/l): peptone, 1; NaCl, 8.5;
sterilized at 115 °C for 20 min.

Preparation of the reference materials

The test strains *Escherichia coli* (WR1), *Enterobacter
cloacae* (WR3) and *Enterococcus faecium* (WR63)[3] were
cultured in a broth, centrifuged and resuspended in 3
litres sterilised evaporated milk. After mixing, the milk
was spray-dried. The contamination level of the resulting
highly contaminated milk powder was 10^5-10^7 colony forming
units (cfu)/ g powder. After storage at 5 °C, these milk
powders were mixed with sterile (γ-irradiated with a dose
of 10 kGy) milk powder to a final contamination level of
2000-3000 cfu/g. The mixing of the powders was done in one
step and later in multiple steps (each step at a ratio
1:1). Gelatin capsules were filled with 0.2-0.3 g mixed
powder, resulting in reference materials with a
contamination level of 400-900 cfu/capsule. Further
details can be found in Mooijman et al[4].

Enumeration

One capsule was reconstituted in 10 ml PS at 38-39
°C, by mixing every 10 minutes up to 40 minutes. After
reconstitution, the suspension was quickly cooled in
melting ice. One ml of the suspension was added to 10 ml
sterile PS in a membrane filtration funnel and filtered
through a membrane filter (0.45 μm pore size). The filter
was incubated on an appropriate selective medium (LSA for
Escherichia coli and *Enterobacter cloacae*, KFA for
Enterococcus faecium). The number of characteristic
colonies on each filter was counted. The counts were
given as number of colony forming units (cfu) per
analytical portion. Here, an analytical portion is defined
as a volume of (1.00 ± 0.02) ml from 10 ml PS in which one
capsule has been reconstituted.

Homogeneity

The variation in the number of cfu between samples of
single reconstituted capsules (duplicate variation,
related to proper mixing of the capsule solution) and the
variation between samples of different capsules of one
batch (related to proper mixing of the powder) were
analysed separately. The theoretical optimal distribution
of microorganisms in the suspensions obtained by thorough
mixing is the Poisson distribution. Two statistical tests
were applied to determine the variation of the number of

cfu within one capsule (test statistic T_1, formula 1) and between capsules of one batch (test statistic T_2, formula 2)[4].

$$T_1 = \sum_i \sum_j [(y_{ij} - y_{i+}/J)^2 / (y_{i+}/J)] \qquad (1)$$

$$T_2 = \sum_i [(y_{i+} - y_{++}/I)^2 / (y_{++}/I)] \qquad (2)$$

Where y_{ij} is the result of subsample j of capsule i, $y_{i+} = \Sigma\, y_{ij}$ (total count of one capsule), J is the number of subsamples of one capsule, I is the number of capsules and $y_{++} = \Sigma\, y_{i+}$ (total count of one batch of capsules).

In case of a Poisson distribution T_1 and T_2 follow a χ^2-distribution with respectively $I(J-1)$ and $I-1$ degrees of freedom. In that case the following ratios: $T_1/\{I(J-1)\}$ and $T_2/(I-1)$, are expected to be equal to 1.

Stability

Long-term stability test. Thirty capsules of batch LWL-050402 and 20 capsules of batch LWL-010403, both with test strain *Enterobacter cloacae* (WR3), were stored at 5 °C and counted monthly (single samples for LWL-050402 and duplicate samples for LWL-010403) for at least half a year.
Challenge test. Forty capsules of batch LWL-141802 with test strain *Enterococcus faecium* (WR63) were stored at -20, 22, 30 and 37 °C. Five capsules of each temperature were analysed in duplicate (on KFA) on day 0, 3, 7, 10, 14, 17, 21 and 31. The results were log-transformed and analysed with linear regression[5].

REFERENCES

3. A.H. Havelaar, W.M. Hogeboom, et al, J. appl. Bacteriol.,1987, 62, 555.
4. K.A. Mooijman, P.H. in 't Veld, et al,'Development of microbiological reference materials', Commission of the European Communities, Luxembourg, EUR 14375EN, 1992, 24 and 30.
5. N.R. Draper and H. Smith, 'Applied regression analysis', J. Willey and Sons, New York, 1966.

3 RESULTS AND DISCUSSION

Homogeneity

It was shown in many different tests that the results of the T_1 test (duplicate variation), always followed the χ^2-distribution. Hence the variation in the number of cfu

in one capsule solution followed the Poisson distribution.
However, the variation in the number of cfu between
different capsules of one batch showed always
overdispersion in comparison with the Poisson
distribution. Most of the early prepared batches showed
much overdispersion (mixed in only one step). The later
prepared batches, which were mixed in multiple steps,
showed better results. However, in practice it seemed to
be impossible to achieve the Poisson distribution.
Furthermore it was also shown by a theoretical approach
that overdispersion is the logical consequence of the fact
that one grain of milk powder may contain more than one
viable bacterial cell[6]. The (in)homogeneity which is
presently accepted is $T_2/(I-1) \leq 2$.

In Figure 1 the results of the homogeneity tests of
batch LWL-141802 with test strain *Enterococcus faecium*
(WR63) are given. This batch was stored at -20 °C and 20
capsules were enumerated in duplicate (on KFA) every
month. This figure clearly shows that for this batch the
results of $T_2/(I-1)$ never exceeded the maximum accepted
value of 2 and that no trend with time could be detected.

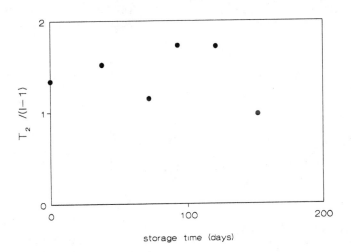

Figure 1 Results of homogeneity test of batch LWL-141802,
with test strain *Enterococcus faecium* (WR63) (I=20),
determined on KFA.

Stability

The stability of the reference materials is
influenced by three factors: a) Age of the highly
contaminated milk powder; b) Test strain and c)
Temperature of storage.
    a) Age of the highly contaminated milk powder. The
dependence of the age of the highly contaminated milk
powder is demonstrated in Figure 2.

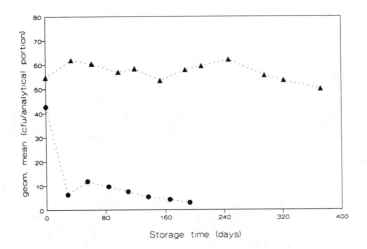

Figure 2 Long-term stability tests at 5 °C of reference materials with test strain *Enterobacter cloacae* (WR3). Batches LWL-050402 (..●..) and LWL-010403 (..▲..) were prepared 1 month and 27 months, respectively after spray-drying.

Batch LWL-050402 with test strain *Enterobacter cloacae* (WR3) was prepared only one month after preparation of the highly contaminated milk powder. In Figure 2 the stability test at 5 °C shows a very rapid decrease in the mean cfu of the materials during the first month. The long-term stability at 5 °C of batch LWL-010403 with the same test strain (prepared from highly contaminated milk powder with an age of 27 months) shows a better stability. An overall decrease of only 9 % of the mean number of cfu was found with the latter batch after one year of storage. Hence, a longer stabilisation time of the highly contaminated milk powder will, in general, guarantee a better stability of the reference materials.

b) Test strain. The duration of the stabilisation period of the highly contaminated milk powder depends on the test strain. A reference material with test strain *Enterococcus faecium* (WR63) showed good stability already 4 months after spray-drying. In general the materials with Gram positive strains show better stability than materials with Gram negative strains.

c) Temperature of storage. All reference materials prepared until now showed the best stability when stored at -20 °C. Storage at higher temperatures, especially above room temperature, will result in a (rapid) decrease of the mean contamination level. This is demonstrated in Figure 3, in which the regression curves of the challenge test of batch LWL-141802 with *Enterococcus faecium* (WR63) are given.

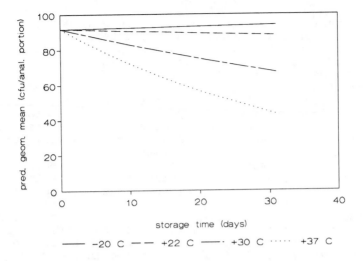

Figure 3 Regression curves of the challenge test on KFA of batch LWL-141802, with test strain *Enterococcus faecium* (WR63).

This batch of reference materials, the best stable material prepared untill now, showed no significant decrease in the mean contamination level after 31 days of storage at 22 °C. Storage at higher temperatures resulted for 30 °C in a slow decrease and for 37 °C in an obvious decrease of the mean contamination level. This limits the possibility for transport.

REFERENCE

6. K.A. Mooijman, P.H. in 't Veld, et al,'Development of microbiological reference materials', Commission of the European Communities, Luxembourg, EUR 14375EN, 1992, 24 and 30.

4 POSSIBLE APPLICATIONS OF THE REFERENCE MATERIALS

A distinction should be made between Reference Materials (RM's) and Certified Reference Materials (CRM's). A reference material can be certified when the parameter of interest can be determined accurately with a so-called definitive method (a method free of any systematic error) in a highly experienced laboratory. More often the parameter is certified on the basis of results obtained by independent methods (methods having no common step in the entire analytical procedure), in experienced and selected laboratories working independently. In chemical analysis such independent method may exist. In microbiology common steps in methods are unavoidable (temperature of growth, culture media, etc.) in order to allow a proper development of the colonies to be counted.

A first certification study has been carried out for a batch of reference materials with test strain *Enterococcus faecium* (WR63). The certified value will be an operational definition of the 'true' value, because a common or similar method(s) has been used for the determinations. Here we can speak about 'methodologically-defined' reference materials, because the certified value is defined by the applied method following a very strict protocol[7].

The certification study of a RM can lead to a high price-level of the CRM and automatically limits their use. Because the contamination level of a microbiological CRM is guaranteed they will especially be useful for special and important quality assessments in a laboratory. Beside the microbiological CRM's it will also be possible to produce RM's by the same process. These RM's will be cheaper than the CRM's but also limited in their use.

The possible applications of microbiological RM's and CRM's include:
a) Testing accuracy in individual laboratories (CRM);
b) Comparing the performance of different laboratories (RM or CRM);
c) Developing and validating methods and media for the detection and enumeration of microorganisms (first with RM, later with CRM);
d) Determining the influence of matrix ingredients and competitive flora on the isolation of a particular organism (first with RM, later with CRM);
e) Providing a standard stable material for collaborative studies (RM);
f) Use in first-line quality control (RM).

The test strains chosen so far for the preparation of the reference materials for water microbiology are strains which respond well to variations in incubation temperatures or methods and media. For instance, the chosen Enterobacteriaceae strains are a thermotolerant (*Escherichia coli*, T_{max}= 45.5 °C on LSA) and a non-thermotolerant (*Enterobacter cloacae*, T_{max}= 42.5 on LSA) strain. When these materials are used for the quality control of the analysis for thermotolerant coliform bacteria, *Escherichia coli* should grow well and *Enterobacter cloacae* should not grow at the prescribed temperature of 44 °C. However, high stacks of Petri dishes, or uneven temperature distributions in the incubator, can easily result in false-positive results, clearly shown by the growth of test strain *Enterobacter cloacae*.

REFERENCE

7. ISO-Guide 35. Certification of reference materials. General and statistical principles. International Organization for Standardization, Geneve, 1989, second edition.

5 CONCLUSIONS

Progress is being made in the development of microbiological reference materials. By optimizing the factors, like age of the highly contaminated milk powder, test strain and mixing procedure, homogeneous and stable materials can be produced. A first certification study has already been carried out. The microbiological reference materials have many possible applications, which are largely influenced by the chosen test strains.

6 ACKNOWLEDGEMENTS

The work was carried out on behalf of the EEC, Community Bureau of Reference (BCR). Dr B. Griepink (BCR), Dr E. Maier (BCR) and R. van der Heide (RIVM) are thanked for their contributions.

Quality Assurance Procedures Schemes for Analysis of Residues in Meat, in the MERCOSUR Countries

Alfredo M. Montes Niño[1] and Oscar D. Rampini[2]

[1]VETERINARIAN, INTERNATIONAL CONSULTANT ON FOOD PRODUCTION, BUENOS AIRES, ARGENTINA
[2]CHEMIST, INTERNATIONAL CONSULTANT ON FOOD ANALYSIS, BUENOS AIRES, ARGENTINA

INTRODUCTION

The MERCOSUR is a common market that will start operating in 1993 between Argentina and Brazil. Paraguay and Uruguay will incorporate in 1995.

The total bovine herd of these four South American countries is of more than 200 million heads.

The particularity of the exports of meat of these countries is that they have followed the most stringent controls of the importing countries' inspections.

As far as Residues are concerned, the application of Quality Assurance has been at constantly increasing levels.

There are more than 20 accredited and official laboratories with a total of more than 150 000 analyses in meat products.

This description of the QA procedures used by these four countries will emphasize the original evolution of these schemes as an example of the possibility of following QA practice in developing countries.

MAIN QUALITY AND QUALITY ASSURANCE CONTROLLED PRACTICES

Residue Laboratories of the MERCOSUR are used to control the following parameters :

a) Detection Limits
b) Quantification Limits
c) Lineal range of the equipment
d) Lineal range of the methods
e) Recovery charts and ranges
f) Critical Control Points
g) Control charts of instruments' parameters, such as:
        "Background noise"
        Gas or liquid flows
            In Immunoassay: specific bound, B50 doses, etc.

h) Resolution of chromatographic columns
i) Training of technicians to reach a standard performance
j) Control of supplies and spare parts
k) Control of samples reception:
        Samples acceptance
        Samples identification
        Time for results output
l) Samples storage
m) Initial processing of the samples.

ARGENTINA

The residues analysis in meat and meat products is the responsibility of the Servicio Nacional de Sanidad Animal (SENASA), authorized institution dependent on the Agriculture Secretariat.

The introduction of quality assurance (QA) practices appeared in 1993 with the necessity of accrediting private laboratories for the analysis of organochlorine pesticides, which had been developed by official laboratories since 1968.

The amount of these analyses, that today total more than 100 000 per year, obliged the government to accredit private laboratories, as it exceeded widely its own analytical capacity, and consequently a control system had to be developed.

This system should allow official inspectors to evaluate the efficiency and accomplishment of 13 accredited laboratories.

It was not named initially QA but it was statistically based; it defined critical points, decision rules, intralaboratories and interlaboratories samples with rankings and corrective actions.

The visits of experts from the Food Safety and Inspection Service of the United States of America (FSIS-USA) introduced the concept of Quality Assurance and the adaptation of the practices executed in Argentina to the Manual of QA of the FSIS. This stage emphasized the registration process, and new forms were designed, specific registers for equipment maintenance, solutions preparations, etc.

Another important step was in 1988 the incorporation of computers in the Official Laboratory. At the moment all the QA on the residues analysis is informatized.

It could be concluded that the decision to accredit private residue laboratories led to a more developed control system, with a more extensive incorporation of chemists and also statistics experts.

The expertise generated by the continuous process of controlling these credited laboratories made it easier to
perform more stringent controls from the meat importers
countries such as the USA and the European Communities.

The more recent initiative was the design of a QA Program
for the whole Laboratory of SENASA,

BRAZIL

The responsibility of the residues analysis in animal
products belongs to the Divisao do Laboratorio Animal,
Secretaria de Defesa Agropecuaria, Ministerio da Agricul-
tura e Reforma Agraria.

Brazil started its residues analysis in organochlorine
pesticides in 1968 but in lower quantities than
Argentina.

Brazilian experts had visited FSIS laboratories and most
of the criteria applied are inspired in the QA Handbook
of the FSIS but local experience has been developing;
from then an independent point of view is expected to
be appearing soon.

The QA Program was seriously redesigned in 1989, and the
advantage of having three official residue laboratories
permitted a more accurate system.

The actual QA program includes a national QA supervisor
and one QA coordinator in each of the three laboratories.

There are defined processes for validation of the new
analytical techniques, definition of critical control
points, defined processes for the incorporation of new
technicians, detailed controls such as temperature controls of refrigerators and freezers.

A regularly updated manual has been produced and this QA
methodology is being extended to all the analytical process of the Divisao do Laboratorio Animal (Animal Laboratory Division) of Brazil.

PARAGUAY

The execution of residues analysis was originally encharged to the Instituto de Tecnologia y Normalizacion.

In 1988, due to the increasing quality demands of the
meat importing countries, the Sub-Secretaria de Gana-
deria (Under-Secretariat of Livestock), which is responsible for the control of animal products, decided to
build a Residue Laboratory.

In the meantime the analyses were performed in official
Argentine laboratories and since 1990 in the Veterinary
Faculty of Asuncion.

The new Laboratory is expected to be operating in March 1993.

An interesting observation is on the fact that all the analytical know-how on residues analysis established in Paraguay since 1990 was transferred from South America.

The QA practices are still in an initial stage, and there is not yet a formal handbook. But QA charts have been developed for most of the analysis, particularly radioimmunoassay.

URUGUAY

The analysis of residues in meat products is the responsibility of the Direccion General de Servicios Veterinarios of the Ministry of Livestock and Agriculture.

The laboratory encharged is the "M.C. Rubino", which has performed the analysis since 1968. A new building was occupied in 1990.

The Residue Program in Uruguay is, in my opinion, the best organized of the area. All the farms are registered and have files with their history. The sampling is based in the history of the farms suppliers of the animal to be slaughtered.

A national commission is encharged with the follow-up of the program and proposes modifications to the regulations or new ones.

Having only one laboratory, the QA Program is based on intralaboratorial controls. The practices are also inspired from the FSIS Handbook.

CONCLUSION

The difficulties that these countries have to deal with are considerable: official budget cuts, great distance from the high technology centers, small markets that do not help to generate suppliers of sophisticated chemical products nor servicing of scientific equipment, and few research activities in these particular fields.

Despite these problems, a considerable advance has taken place in the establishing of QA programs for the analysis of residues.

These programs are strongly oriented to fulfil the requirements of importing countries, but it has to be admitted, on the other side, that this part of the world is one of the less contaminated and consequently national health priorities are others by far.

Another interesting item to note is that in the area

QA programs are not very common, either in private or in public institutions. And the quality performance of the residues analysis is considerably higher than other chemical activities.

To appreciate the consequences of these activities, we can mention for example that in Argentina residues expert chemists participate actively in the quality commission of the national standards institution, IRAM, which is affiliated to ISO.

Translation of AOAC Methods for Use by Technicians in the Laboratory to Obtain Quality in Their Analysis

David N. Holcomb

SILLIKER LABORATORIES, 1304 HALSTED STREET, CHICAGO HEIGHTS, IL 60411, USA

Introduction

The Official Methods of the Association of Official Analytical Chemists (AOAC) [1] are often stipulated by governments as well as commercial contracts. These excellent methods are written in a very condensed form and thereby many methods can fit into convenient, easy-to-handle books. While the AOAC methods are complete and well-written, we find that technicians may have difficulty understanding and following them because of the extensive abbreviation and cross-referencing needed to make the methods sufficiently concise to fit into the two Official Methods volumes. If a technician has difficulty following a method, the quality of results that he/she generates may be in question. Therefore, expanded versions of these methods have been prepared for use by Technicians in the Laboratory.

Example -Ash by Ignition

An example of an expanded version of a method is given in this paper. The method chosen for illustration is the AOAC Official Method 935.42, "Ash of Cheese: Gravimetric Method" [1]. The Official method is based on reports by Stone [2, 3] which were published over 55 years ago, but it is still regularly referenced [4,5] as the method of choice for ash determination. The Official Procedure is reprinted, then the expanded procedure follows. Although a degreed chemist would have no difficulty carrying out the determination with the AOAC method, a non-degreed technician will find the expanded version helpful.

The expanded method is printed using a common word processing program (Word-for-Windows). The variation in type size gives the document a "professional" appearance which reinforces the "quality" concept to the technician. The title page includes a procedure number following a system already in use in the laboratory

The first page of the method (after the title page) is a table of contents which gives the layout of the method. The overview (section 1) includes a brief description and a statement of the scope or applicability of the procedure. The first section also includes "Procedure Referenced By", "Procedure References", "Worksheet Heading" and "Report Heading" sub-sections. These sub-sections are used within the laboratory to help the technician identify the correct forms to use for data recording and reporting.

A flow diagram follows the Overview section. The graphic display of the flow diagram is preferred by some technicians who find it easier to follow than ordinary textual material. The flow chart was prepared using "Flow Charting 3", Patton & Patton Software Corporation, Morgan Hill, CA.

The highlights section (section 3) provides the technician a brief working document which can be laminated in plastic and used at the bench. For a relatively brief method such as Ash Determination, the "Highlights" section is not much shorter than the full procedure (section 8), but in most cases, it summarizes a more comple set of instructions.

Section 10 (Special Notes) is used to emphasize procedural steps which are especially important to successful completion of a determination. The quality assurance section (Section 11) gives values to be expected from check samples, may include statements about precision and accuracy, and again may emphasize steps which are critical to a "quality" determination. The reference section of the method is expanded to include more recent references if those are available and relevant text book references. Safety notes specific to the method are included in Section 13. Part of the auditing procedure is to insure that the technician performs the procedure in a safe manner. The final section of the method provides space fo signatures and approvals. These should be entered by management level people to add emphasis to the importance of following the procedure without deviation.

The rest of the expanded version follows the outline given in the Table of Content and is similar to the AOAC "Guide to Method Format" [1].

In addition to the text, a checklist and a flow diagram have been included.   The checklist serves both as a reminder to the technician to see that critical steps are carried out correctly and as a training aid to the supervisor. It can also be used for internal auditing, as in preparation for an ISO registration audit. The flow chart serves those same purposes, but presents the procedure in a more graphic format which is preferred by some technicians.

Issues

An issue with any revision of an <u>Official Method</u> is the question of whether or not the revision faithfully reproduces the original method. Careful review of the method by several people who are familiar with the AOAC procedure must be done. A technician who follows the expanded procedure must obtain the correct results on daily check samples and NIST standard reference materials. Satisfactory performance in accreditation programs (such as that of USDA) and collaborative studies indicates that the analytical system is in control and that the procedural steps are the correct ones.

References

1. K. Helrich (ed.). "Official Methods of Analysis of the Association of Official Analytical Chemists, 15th edition", Association of Official Analytical Chemists, Inc. Arlington, VA, 1990. [See several methods in this reference for ash determination on different foods.]

2. C. B. Stone, <u>J. Assoc. Off. Analyt. Chem.</u>, 1935, <u>18</u>, 401.

3. C. B. Stone, <u>J. Assoc. Off. Analyt. Chem.</u>, 1937, <u>20</u>, 339. 339-341.

4. Y. Pomeranz and C. E. Meloan, "Food Analysis: Theory and Practice, Second Ed.", AVI-Van Nostrand Reinhold Company Inc., New York. 1987, Chapter 34, p. 612.

5. R. S.Kirk and R. Sawyer, "Pearson's Composition and Analysis of Foods, Ninth Ed.", Longman Scientific & Technical, England, 1991.

Illustration 1: AOAC Official Method

935.42   Ash of Cheese
                   Gravimetric Method
                       Final Action

Weigh 3–5 g prepd sample, **955.30**, into Pt dish, place on steam bath, and dry ca 1 hr. (If cheese is high in fat, place small amt of absorbent cotton in dish.) Ignite cautiously to avoid spattering and remove burner while fat is burning. When flame ceases, complete ignition in furnace at $\leq 550°$, cool, and weigh.

Refs.: JAOAC **18**, 401(1935); **20**, 339(1937).

Illustration 2: Expansion of Ash Method (pp. 136-143)

0225-2001 ASH BY IGNITION (DRY ASHING)
Revision: 01/14/93
Supercedes: 09/16/93
Page 2 of 10

1.   OVERVIEW

BACKGROUND AND SCOPE

A known weight of sample is heated above its ignition point (to 550°C) and held at that temperature overnight. All organic compounds are oxidized to carbon dioxide, carbon monoxide and water. The remaining carbon-free residue is weighed and the result is expressed as percent ash. Nearly all food products can be analyzed by this method.
The wet ash method, which is used primarily for the digestion of samples for determining trace elements and metallic poisons, is given as a separate procedure.

PROCEDURE REFERENCED BY

PROCEDURE REFERENCES

0225-2001 ASH BY IGNITION (DRY ASHING)
Revision: 01/14/93
Supercedes: 09/16/92
Page 3 of 10

1.   OVERVIEW (continued)

WORKSHEET HEADING

ASH %

REPORT HEADING

ASH %

ASH BY IGNITION (DRY ASHING)

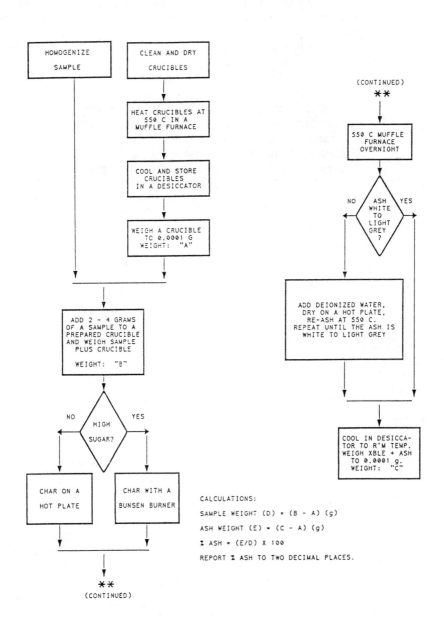

CALCULATIONS:

SAMPLE WEIGHT (D) = (B - A) (g)

ASH WEIGHT (E) = (C - A) (g)

% ASH = (E/D) X 100

REPORT % ASH TO TWO DECIMAL PLACES.

3.   HIGHLIGHTS

1. All weighings should be made and recorded to 0.1 mg (0.0001g).
2. Two-to-four grams of sample is placed in a clean, dry crucible.
3. Prior to ignition in a muffle furnace, the sample is charred on a hot plate. (High sugar products - such as candies- are charred with a Bunsen burner.)
4. The charred sample is ignited in a muffle furnace at 500 - 550°C. (See the Full Procedure for comments regarding temperature of the muffle furnace.)
5. After ignition, the ash should be white or light grey. A dark grey color indicates incomplete ashing which would lead to overestimation of the ash content.
6. The weight of ash is determined.
7. The ash content is reported as percent by weight.

4. TRAINING AND AUDITING CHECKLIST

Name: _____ Reviewer: _____ Date: _____

Control Point	Satisfactory (S) or Unsatisfactory (U)	Comments
Sample Properly Mixed or Blended		
Crucibles Thoroughly Clean and Dry		
Sample Properly Charred		
Proper Use of Fume Hood		
Muffle Furnace Temperature		
Ash White or Light Grey		
Crucible Cooled Before Weighing		
Calculations		

5.   APPARATUS

1.   Balance - Analytical, with 0.0001 g sensitivity.
2.   Crucibles - porcelain, 15 mL or larger (See special note).
3.   Hot plate
4.   Tongs - 18" (min.) length.
5.   Muffle Furnace - 550°C. See the table in "Special Notes" (Section VII) for other recommended temperatures.
6.   Desiccator - With desiccant.
7.   Fume Hood.
8.   Bunsen Burner - For charring very high sugar samples.

6.   REAGENTS

Water - Deionized..

7.   SOLUTIONS AND MIXTURES

None.

8. FULL PROCEDURE

1. Preparation of crucibles:
   a. Thoroughly clean and dry the crucibles.
   b. Heat the crucibles in the 550 deg.C muffle furnace.
   c. Remove the crucibles from the muffle furnace with tongs. Cool and store the crucibles in a desiccator until they are to be used.
2. Weigh a prepared crucible to 0.0001 g and record the weight.
3. Add a sample to the crucible. Usually use a 2 - 4 g sample, but refer to the AOAC (1990) methods for specified weights for different products. Weigh the crucible plus sample to 0.0001g.
4. Gently heat the sample in the crucible on a hot plate in the fume hood to drive off moisture.
5. Increase the temperature slowly. As the sample begins to char, smoke will be liberated. As the smoke decreases, raise the temperature on the hot plate and continue this process until no more smoke is given off from the sample at the highest setting on the hot plate.
5 a. For very high sugar samples (e.g., pure sugar, some high sugar candies), the sample should be charred with a Bunsen burner.
6. Place the sample in the 550 deg. C muffle furnace overnight.
7. Remove the crucible containing the sample's ash from the muffle furnace and cool it in a desiccator. The residue should be white or light grey in color.
8. When the crucible has cooled to room temperature, weigh and record the weight as "Weight: crucible + ash".

9. CALCULATIONS AND REPORTING

Sample Weight = [(Wt. crucible + sample) - (Wt. crucible)] (grams)

Ash Weight = [(Wt. crucible + ash) - (Wt. crucible)] (g)

% Ash in Sample = [(Ash Weight)/(Sample Weight)] X 100

10. SPECIAL NOTES

1. Handle prepared crucibles (V.1) with tongs. <u>Never</u> handle prepared crucibles by hand as oils from the hands may cause contamination, leading to high ash values.

2. When heating a sample on the hot plate, increase the temperature <u>slowly</u>. Avoid spattering which could cause sample loss, leading to a low ash result.

3. Although 550°C is a satisfactory ashing temperature for most foods, some products may require different ashing temperatures in the muffle furnace. The following table was extracted from the AOAC Official Methods [see the Helrich (1990) reference].

Product	AOAC Method No.	Temperature (°C)
Baking Powder	920.46	500
Butter	920.117	500 (max.)
Fruit Products	940.25	525 (max.)
Honey	920.181	600
Maple Sirip	920.187	600

4. If the incineration process does not yield a white or light grey residue, the ashing process may not be complete, and high ash results may be obtained. Moisten the sample with deionized water, re-dry it on the hot plate and put it back in the muffle furnace.

5. Products high in carbohydrates may require a larger crucible to prevent loss of sample due to foaming.

11. QUALITY ASSURANCE

Ham is used as an internal check sample. The current (01/93) check sample has an ash content of 5.28 ±0.12 percent.

0225-2001 ASH BY IGNITION (DRY ASHING)
Revision: 01/14/93
Supercedes: 09/16/92
Page 10 of 10

12. REFERENCES

Helrich K. (ed., 1990). Official Methods of Analysis of the Association of Official Analytical Chemists, 15th edition. Association of Official Analytical Chemists, Inc. Arlington, VA. [See several methods in this reference for ash determination on different foods.]

Kirk, R. S. and Sawyer, R. (1991). Pearson's Composition and Analysis of Foods, 9th Edition. Longman Scientific & Technical, England.

Pomeranz Y, Meloan CE. (1987). "Ash and Minerals", Chapter 34 in: Food Analysis: Theory and Practice, Second Edition. AVI-Van Nostrand Reinhold Company Inc., New York. 612-635.

Stone, C. B. (1935). "Report on Cheese", J. Assoc. Off. Analyt. Chem., 18, 401-402.

Stone, C. B. (1937). "Report on Cheese", J. Assoc. Off. Analyt. Chem., 20, 339-341.

13. SAFETY

Safety glasses, gloves and a laboratory apron must be worn while doing this procedure. Gases given off in the course of this procedure may be toxic and the ignition process should be done in a fume hood with an air flow of at least 100 cubic feet per minute. The muffle furnace is extremely hot (550°C) and so are the crucibles which have been placed in it. Severe burns could result from touching or handling the hot items. Long (at least 18") tongs should be used in handling the crucibles. Safety glasses should be worn while doing this procedure. **IF YOU HAVE ANY QUESTIONS REGARDING THIS PROCEDURE, INCLUDING HOW TO PERFORM IT SAFELY, CONTACT YOUR SUPERVISOR.**

14. SIGNATURES AND APPROVALS

Prepared by: _____   Date: _____

Approved by: _____   Date: _____

Toxic Trace Elements in Chilean Seafoods: Development of Analytical Quality Control Procedures

I. Do Gregori[1], D. Delgado[1], H. Pinochet[1], N. Gras[2],
L. Muñoz[2], C. Bruhn[3], and G. Navarrete[3]

[1]CHEMICAL INSTITUTE, CATHOLIC UNIVERSITY OF VALPARAISO, CASILLA
4059, VALPARAISO, CHILE
[2]NEUTRON ACTIVATION ANALYSIS LABORATORY, CHILEAN ENERGY
COMMISSION FOR NUCLEAR ENERGY, CASILLA 188D, SANTIAGO, CHILE
[3]INSTRUMENTAL ANALYSIS DEPARTMENT, PHARMACY FACULTY,
UNIVERSITY OF CONCEPTIÓN, CASILLA 237, CONCEPTIÓN, CHILE

1 INTRODUCTION

Chile is a country with approximately 4500 kilometers of continental coastline of the South Pacific Ocean, with diverse geographical zones and with ocean waters of different abiotic and biotic characteristics. It is therefore in a favorable position to develop fishing activities, since its waters contain a great variety of marine resources, namely, fish, shellfish and seaweeds. Fishing in Chile plays an important role. In 1988 alone export volumes reached a total of over 5 million tons[1], thus placing Chile as one of the important fish exporting countries in the world.

Environmental pollution is a growing hazard to human health. Worldwide, the chemical industries are proliferating and their harmful effects on the environment are increasing over the years[2]. As harmful environmental pollutants, heavy metals are the second most important toxicants causing adverse effects on organisms, and have caused concern in the marine environment. Therefore, sea food consuming countries have gradually increased their surveillance regarding maximum levels of metal contamination allowed for these products, as reflected by stricter quality control approaches for seafoods.

In turn, these developments have increased the need for the exporting countries to enhance the quality of their products to comply with prescribed standards. In any case adequate quality assurance is required for every environmental investigation involving analyses of biological materials, probably the most crucial point being reliable sampling and sample treatment.[3,4]

Taking all the facts into consideration, three Chilean Research Institutions, joined together to develop the analytical procedures. These institutions had the required expertise for determining toxic trace elements by at least 3 different analytical techniques, Neutron Activation Analysis (NAA)[5], Anodic and Cathodic Stripping Voltametry (DPASV)[6,7] and Atomic Absorption Spectroscopy (AAS)[8,9]. The objective of this collaboration was to establish sampling procedures and protocols, sample preparation and treatment, and the application of qualified instrumental methods for the analysis of seafoods, such as razor clams, oysters, clams, pink clams, mussels etc. The potential of shellfish analysis for monitoring human health risks from toxic elements like Cd. Pb and Hg is well documented.[10-12]

This paper describes different sample treatment procedures (including sampling, dissection, homogenization, lyophilization and dissolution) that have been experimented in the analyses of trace element in different chilean seafoods, and analysis by the three different techniques. In order to assure quality control and assessment, different reference materials were analyzed and the analysis of real samples were not undertaken until the analytical performances were satisfactory. In connection with these studies, interlaboratory comparisons were also realized.

2 EXPERIMENTAL

Sampling and Samples Storage

Samples of fresh mollusc (*Semelle Solida and Tagelus Dombeii*, "Almejas" and "Navajuelas Chilenas" respectively) were manually caught by a diver from three different areas of the Chilean coast, well known as natural commercial bank of these species: Arauco gulf (VIII region), Corral Bay and Ancud Gulf (X region), and also at the initial stages of seafood export operations at the canning industry.

All samples were collected in precleaned vessels and stored deep frozen in precleaned plastics bags. The specimens were classified according to their size.

Sample Preparation

Preparation and treatment of all samples to be analysed was carried out in adequate working areas. Two interconnected rooms were equipped for this purpose at the Nuclear Research Centre La Reina, near Santiago. The first room is a "preclean area" designated for change of clothes and initial storage of samples. The second room in the "clean area" (that had a laminar fumehood) was used for sample treatment. The walls of the Laboratory and ceiling are masked with white epoxy paint, and the floor is covered with vinyl and the doors and windows are completely sealed with special materials, thus avoiding external contamination.[13]

Mollusc samples were thawed for a few minutes prior to analysis in a microwave oven, then opened with a titanium knife and the tissue separated from the shell. Each sample was divided into two subsamples, one of them was dissected and the visceral tissue was separated from the muscular mass. Samples were crushed and homogenized using a plastic food processor (Moulinex), specially adapted with a high purity titanium blade which was ingeniously designed and fabricated. Homogenized samples were transferred to special plastic flasks and were lyophilized for 72 hours at 12°C and at 0.1 mbarns pressure. The moisture loss during this process was determined for all samples. Lyophilized samples were homogenized again and preserved in desiccators at room temperature until analysis was carried out. Sample humidity was determined by heating a known sample mass at 100°C for 16 hours. Results refer to dry mass.

All materials for sampling and sample treatment were acid cleaned, rinsed with adequately pure water and stored in polyethylene bags until used. All handling operations were carried out using plastic implements, quartz or titanium blades and stored with great care in order to prevent contamination risks. Reagents for treatments, digestions and analytical determination were of ultrapure quality.

3 ANALYTICAL METHODS

For the determination of cadmium were used anodic stripping voltammetry, atomic absorption spectrometry and radiochemical neutron activation analysis as analytical techniques. Mercury was determined by atomic absorption spectrometry and instrumental neutron activation analysis. For copper determinations were used radiochemical neutron activation analysis and atomic absorption spectrometry.

The analytical methodology and instrumental conditions employed by the different techniques are recently described in detail by the authors.[14]

4 RESULTS AND DISCUSSION

Analysis of Certified Reference Material

Monitoring the metal bioaccumulation in organisms needs a strict quality control of analysis. Proper application of quality assurance requires the analysis of Certified Reference Materials (CRMs) that matches as closely as possible with matrix type and element level of real samples.[9,10] For this purpose, the use of working standards, the extended participation in interlaboratory intercomparison and where possible, calibration against appropiate standard reference materials, become necessary.[15,16] One difficulty in this approach is the lack of an appropiate Certified Reference Material (CRM) with a similar matrix as the mollusc object of this study and also, with the desirable concentration levels of heavy metals. Therefore to obviate this situation, we use various NIST, IAEA and BCR "CRM", with different concentrations of these toxic metals, for the analytical quality control. Some of the results obtained for Certified Reference Materials analyzed by different techniques are summarized an Table 1.

Actually it is known that cadmium concentrations in canned mussels from Chile (Navajuelas Chilenas) is higher than that found in European mussels.[17] Since, there are no CRM for this mussel, the Institute of Applied Physical Chemistry, Research Centre of Julich, prepared a lyophilized homogeneized powder of Navajuelas Chilenas. This material was distributed, for assessment the cadmium content, among 22 different laboratories from Chile, Spain, Austria, Czechoslovakia and the Federal Republic of Germany.[18]

Table 1.
Results obtainded for the CRM analyzed by NAA, DPASV and AAS. (Values: ug g⁻¹ ± CL, dry wt)

CRM	Cd			Cu			Hg		
	Obtained	Certified or Recommended	%Diff.	Obtained	Certified or Recommended	%Diff.	Obtained	Certified or Recommended	%Diff.
Oyster Tissue NBS 1566	3.4±0.5 3.5±0.2 3.6±0.4	3.5±0.5 3.5±0.5 3.5±0.5	-2.9 (1) -2.9 (2) -2.9 (3)	62±3 60.0±0.9	63.0±3.5 63.0±3.5	-1.6[1] -4.1[2]	0.032±0.003 0.060±0.007	0.057±0.015 0.057±0.015	-44[1] 5.2[2]
Orchard Leaves NBS 1571	<0.65 0.11±0.03	0.11±0.01 0.11±0.01	[1] 0[3]	12.2±0.2	12.0±1.0	1.7[1]	0.15±0.03	0.155±0.01	-5.0[1]
Copepoda MA-A1/TM AIEA	<1.9 0.7±0.1	0.75±0.03 0.75±0.03	[1] -2.7[3]	7±1	7.6±0.2	-4.0[1]	0.20±0.01 0.30±0.02	0.20±0.01 0.20±0.01	0[1] 6.4[2]
Estuarine Sediment SAM 1646	0.37±0.08	0.36±0.07	2.8[3]				0.077±0.002	0.063±0.012	22[2]
Navajuela Chilena	7.3±0.3 7.6±0.2 7.4±0.5	7.4±0.3 7.4±0.3 7.4±0.3	-1.3[1] 2.7[2] 0[3]	5.9±0.6 5.4±0.4	** **	[1] [2]	0.05±0.01 0.07±0.01	** **	[1] [2]

(1) : NAA
(2) : AAS
(3) : DPASV

** : No recommended values available

Table 2.
Cd. Cu and Hg content in fresh and canned molluscs (Values: ug g^{-1} . dry wt)

SPECIMEN	Cd AAS	Cd NAA	Cd DPASV	Cu AAS	Cu NAA	Hg AAS	Hg NAA
FRESCH NAVAJUELA CHILENA	4.6±0.7						
ARAUCO GULF CMV		5.2±1.5	4.0±0.6	2.4±0.2	2.5±0.2	0.09±0.01	0.14±0.03
CANNED NAVAJUELA CHILENA							
ARAUCO GULF SMV	2.1±0.1	2.0±0.9	2.0±0.4	1.8±0.1	2.1±0.4	0.11±0.01	0.10±0.02
FRESH NAVAJUELA CHILENA							
CORRAL BAY CMV	3.4±0.6	4.1±0.4	3.4±0.6	2.9±0.5	2.4±0.3	0.04±0.01	0.06±0.01
CANNED NAVAJUELA CHILENA							
CORRAL BAY SMV	2.5±0.1	2.4±0.7	2.3±0.9	2.0±0.1	2.0±0.2	0.04±0.1	0.05±0.01
CANNED ALMEJA CHILENA							
ANCUD GULF	1.6±0.1	1.7±0.5	1.7±0.3	3.9±0.1	4.4±0.2	0.03±0.01	< 0.03

SMV: Without visceral tissue.
CMV: With visceral tissue.

Homogeneity Test

In order to determine the homogeneity of samples, 40 kg of fresh mollusc were processed[14]; the variations of Cd and Cu contents in four randomly selected flasks of a same sample were determinded by DPASV and NAA, respectively.

Analysis of Fresh and Canned Molluscs

Based on the consistency of the intercomparison results obtained by the participant laborarories, different fresh and canned molluscs samples (two types) caught in three different geographical locations (Ancud Gulf, Corral Bay and Arauco Gulf) of the Chilean continental coast which are of economic interest, were analyzed for Cd, Cu and Hg content. Some results obtained are shown in Table 2. As can be seen, the results obtained by three different techniques are quite similar and do not show statistically significant differences. This consistency is valid in a wide concentration range and in different mollusc matrices.

The investigation included analyses of samples with and without the visceral tissue, as it was suspected that these two compartments may differ in their toxic metal profile. Indeed, this turned out to be an important factor especially for cadmium. This led to the recommendation of eliminating visceral tissues from the final products as a solution to the industrial problem.

The results of this field assay point out that it is important now to continue this study to establish the sources (e.g. water, sediment) of these metal contents in the molluscs. Such on-going monitoring activities will be helpful to identify the locations of economic importance to the Chilean Seafood Industry.

REFERENCES

1. Fishering National Services of Chile (SERNAP), "Anuario Estadistico", Ministerio de Economia Fomento y Reconstruccion,1988.
2. WHO Report of a WHO Expert Committee, Techn Rep. Ser, 1985, p 718.
3. H.K. Lawrence, W. Crummet, J. Deegan, R. A. Libby, J.K. Taylor, *J. Anal Chem.*, 1983,*55*,2210.
4. G.V. Iyengar, "Elemental Analysis of Biological Systems", CRC Press, USA, 1989, Vol. I.
5. K. Heydorn, "Neutron Activation Analysis for Clinical Trace Element Research", CRC Press, USA, 1984, Vol I.
6. P. Ostapczuk, P. Valenta, H. Rutzel and H.W. Nurnberg, *Sci., Total Environ.*, 1987,*60*,16.
7. H.W. Nurnberg, *Pure and Appl. Chem.*, 1982,*4*,853.
8. K.R. Sperling, *Fresenius Z. Anal. Chem.*, 1988,*332*,565.
9. R. Blust, A. van der Linder, E. Verheyren and W. Declair, *J. Anal. At. Spectrom.*, 1988,*3*,387.
10. D. F. Roberts, M. Elliot and P. A. Read, *Marine Environ. Res.*, 1986,*18*,165.
11. D. J. H. Phillips, *A. Review Environ. Pollut.*, 1977,*13*,281.

12. M. Gordon, G. A. Knauer and J. H. Martin, *Mar. Pollut. Bull.*, 1980,*11*,195.
13. J. R. Moody, *Anal. Chem.*, 1986,*7*,257.
14. I De Gregori, D. Delgado, H. Pinochet, N. Gras, L. Muñoz, C. Bruhn and G. Navarrete, *Sci., Total Environ.*, 1992,*111*,201
15. V. Iyengar, W. R. Wolf and J. Tanner, *Fresenius Z Anal. Chem.*, 1988,*332*, 549.
16. M. Inhat, *Fresenius Z Anal. Chem.*, 1988,*332*,568.
17. P. Ostapczuk and M. Froning, *Fresenius Z Anal. Chem.*, 1990,*337*,104.
18. F. Queirolo, P. Ostapczuk, P. Valenta, S. Stegen, C. Marin, F. Vinagre and A. Sanchez, *Determinación de cadmio en Tagelus Dombeii (Navajuelas) mediante voltametría* de redisolución anódica (DPASV). Proceedings of IX Congreso Latinoamericano de Electroquíca, Tenerife España, 1990, p 394

Philippine Interlaboratory Test Program on Proximate and Mineral Analyses of Foods

T.R. Portugal, A.V. Lontoc, E.M. Castillo, R.S. Sagum,
J.G. Ardena, P.M. Matibag, A.R. Aguinaldo, and
E.M. Avena

FOOD AND NUTRITION RESEARCH INSTITUTE, DEPARTMENT OF SCIENCE
AND TECHNOLOGY, PEDRO GIL COR TAFT AVE., ERMITA 1000, MANILA,
PHILIPPINES

1   INTRODUCTION

The Food and Nutrition Research Institute (FNRI) is the Philippine government agency mandated to lead the conduct of food and nutrition research in the country toward improvement of the nutritional status of Filipinos.

In line with the goal of the International Network of Food Data System (INFOODS) and the ASEAN Food Data Network (ASEANFOODS) to produce complete, accurate, up-to-date, accessible and compatible Food Composition Tables, FNRI initiated the organization among private and government laboratories of a Philippine Interlaboratory Test Program (ITP) on proximate and mineral analyses of foods.

To determine the performance of laboratories analyzing proximate and mineral composition of foods, an ITP was organized among 15 laboratories in Metro Manila and Los Banos, Laguna. Specifically, the study had the following objectives:

1   to determine accuracy of proximate and mineral measurements carried out by laboratories, using NBS mixed diet Standard Reference Material and IRRI in-house rice plant reference material.

2   to compare within- and between-laboratory precision in the analysis of proximate and mineral composition of selected food matrices, using "own" routine analytical procedures.

3   to derive ITP "consensus" methods designed to improve laboratory performance.

4   to assess any improvement of laboratory performance with the use of the "consensus" methods.

2   METHODOLOGY

Design of the Study

The study was conducted in two batches: Batch I
using laboratory's "own" routine methods and Batch II
using the ITP "consensus" methods in the analysis of
moisture, crude protein, crude fat, ash, crude fiber,
iron, phosphorus and calcium.

Participating Laboratories

Fifteen (15) laboratories from two (2) private and
thirteen (13) government institutions including the
academe that are conducting food composition analysis
participated in the study.

Test Materials

Three food materials namely: rice flour and
mungbean flour from agricultural research institutions,
and whole milk powder from a processing company were
used in the study.

Reference Materials

Reference materials were used to evaluate
accuracy/precision of analyses by comparison with
recommended or establihed values namely: NBS mixed diet
SRM 8431 and IRRI in-house rice plant reference
material.

Derivation of ITP Consensus Methods

After evaluation of Batch I ITP results, the
participant   laboratories   agreed   to   adopt
recommendations for improvement of methods within
limits of available resources. Thus "consensus"
methods were either the same, step-by-step throughout
the analysis, or different in some few steps that were
not deemed very critical.

Laboratory Performance Assessment

Means were assessed relative to method-related
variations.

Accuracy was assessed in terms of conformance with
the recommended values for NBS mixed diet reference
material (Batch I), and the established values for the
IRRI in-house rice plant reference material (Batch II).
"Conformers" were defined as those values generated by
individual laboratories that are within range of
recommended values for NBS mixed diet reference
material and the IRRI established values for in-house

rice plant reference material. "Non-conformers" were those values outside the recommended or established values for the reference materials.

For precision assessment, the achievable within-laboratory SD was computed using the formula:

$$SD = \frac{\text{Horwitz achievable CV} \times \overline{X}}{100}$$

The achievable between-laboratory CV was based on Horwitz equation. The FNRI between-laboratory CV cut-off incorporated a 50% increment over the Horwitz CV.

3   RESULTS AND DISCUSSION

The assessment of laboratory performance gave the following results:

Accuracy and Precision of Measurements Carried Out on Reference Materials

NBS Reference Materials – Mean values of all laboratories for iron and calcium (Table 1) did not conform with the recommended values, with high CV´s reported (128.1% and 22%, respectively). For phosphorus, mean values of 3 out of 4 laboratories and

Table 1   Mean (X) assays per 100g (dry basis) of individual laboratories for proximates and minerals of NBS mixed diet reference material

Lab Code	Crude Protein (g)	Crude Fat (g)	Ash (g)	Iron (mg)	Phosphorus (mg)	Calcium (mg)
B	22.30*	10.94*	3.30*	–	–	–
C	22.10*	–	–	–	–	–
G	19.65	11.35*	2.99	4.0*	320	220*
M	19.80	9.92	3.06	–	314	132*
N	19.62	8.88	3.06	n.d*	340	219*
O	18.66	9.39	3.02	0.2*	596*	215*
n	6	5	5	3	4	4
Mean (X̄)	20.4	10.1	3.1	2.1	390	197
CV(%)	7.3	10.3	4.0	128.1	33.4	22.0
NBS Recmd Value (X̄ +SD)	19.1 ±0.62	9.3 ±0.92	3.0 ±0.09	3.7 ±0.26	332 ±21	194 ±14

* Non-conformers relative to NBS recommended values

for ash, 4 out of 5 laboratories were "conformers".
Except for one laboratory that used titrimetric method,
all colorimetric methods used for phosphorus were
applicable to the mixed diet matrix. No method-related
trend was observed for crude protein and ash.

IRRI Rice Plant Reference Material – Similar
observations on accuracy and precision were obtained
for IRRI rice plant material, even with the use of
"consensus" methods.

Precision of Measurements Carried Out on Test Materials

Precision tests indicated large within-laboratory
SD's and between-laboratory CV's in Batch I ("own"
methods) ITP, which decreased only slightly for most
analytes in Batch II ("consensus" methods) ITP,
indicating improvement when analytical protocols were
harmonized.

Results for phosphorus in mungbean flour (Figure
1 showed a high value reported by Lab O in Batch I
using titrimetric method. The "consensus"
colorimetric methods used by all laboratories resulted
to a decrease in within-laboratory SD of Labs D and O
and the between-laboratory CV (5.7%) which conformed
with the FNRI CV cut-off (Table 2). The same trend was
obtained for phosphorus in rice flour and milk powder.

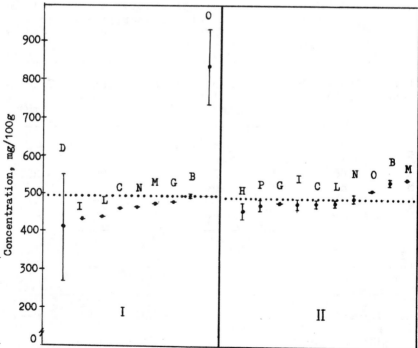

Figure 1   Results of Phosphorus in Mungbean Flour

Table 2   Results of Batch I and II ITP for phosphorus, by test material

Test material	I	II

A.   Rice Flour

	I	II
n	9	10
Individual $\bar{X}$ range	119–428	104–140
W/n-lab SD range	0.71–77.78	0–6.36
Achievable w/n-lab SD	8.63	6.71
Pooled mean ($\bar{X}$)	166.0	122.0
B/w-lab CV (%)	59.8	10.4
Hor. ach. b/w-lab CV(%)	5.2	5.5
FNRI b/w-lab CV cut-off(%)	7.8	7.8

B.   Mungbean Flour

	I	II
n	9	11
Individual $\bar{X}$ range	413–832	455–538
W/n-lab SD range	0–206.48	0.71–22.62
Achievable w/n-lab SD	21.91	21.60
Pooled mean ($\bar{X}$)	498.0	491.0
B/w-lab CV (%)	25.5	5.7
Hor. ach. b/w-lab CV(%)	4.4	4.4
FNRI b/w-lab CV cut-off(%)	6.6	6.6

C.   Milk Powder

	I	II
n	9	10
Individual $\bar{X}$ range	597–1642	720–860
W/n-lab SD range	0–49.49	0.07–9.20
Achievable w/n-lab SD	34.15	32.47
Pooled mean ($\bar{X}$)	833.0	773.0
Between-lab CV(%)	37.0	5.2
Hor.ach. b/w-lab CV(%)	4.1	4.2
FNRI b/w-lab CV cut-off(%)	6.2	6.3

"Consensus" methods that did not result in improvement in within-laboratory SD and between-laboratory CV for the three test materials were observed in ash. This may indicate matrix effects. A slight decrease in CV for milk powder but a tremendous increase in rice flour (Table 3) from "own" methods to "consensus" methods indicated that ashing at 550°C for 2 hr may not be applicable to rice flour matrix. Matrix-applicable methods have to be tested.

Table 3   Results of Batch I and II ITP for asn, by test
          material

Test material	I	II
A. Rice flour		
n	12	12
Individual $\bar{X}$ range	0.32-0.92	0.30-12.04
W/n-lab SD range	0-0.05	0-0.99
Achievable w/n-lab SD	0.02	0.07
Pooled mean ($\bar{X}$)	0.54	1.88
B/w-lab CV (%)	25.9	183.0
Horwitz ach. b/w-lab CV(%)	4.4	3.8
FNRI b/w-lab CV cut-off(%)	6.6	4.4
B. Mungbean Flour		
n	12	12
Individual $\bar{X}$ range	3.68-4.63	3.20-5.50
W/n-lab SD range	0-0.17	0-1.33
Achievable w/n-lab SD	0.13	0.14
Pooled mean ($\bar{X}$)	4.20	4.22
B/w-lab CV(%)	7.1	13.3
Hor. ach. b/w-lab CV (%)	3.2	3.2
FNRI b/w-lab CV cut-off(%)	4.8	4.8
C. Milk Powder		
n	12	12
Individual $\bar{X}$ range	5.33-5.96	5.44-6.04
W/n-lab SD range	0-0.96	0.01-0.25
Achievable w/n-lab SD	0.18	0.18
Pooled mean ($\bar{X}$)	5.76	5.78
Ach. b/w-lab CV(%)	3.1	2.9
Hor. ach. b/w-lab CV (%)	3.1	3.1
FNRI b/w-lab CV cut-off(%)	4.6	4.6

4   CONCLUSION

The ITP's demonstrated the usefulness to analytical
laboratories   of   participating   in   local   and
international   intercollaborative testing program to
ensure good quality data. Local development of food
reference materials need to be initiated for ready
availability and economy. Matrix applicable methods
need to be adopted and their use   harmonized   among
local laboratories within the country for local and
international data interchange.

5   REFERENCES

1.   Horwitz, W. E. 1980. Evaluation   of Analytical
     Methods Used for Regulation of Foods and Drugs,
     Anal. Chem. 54(1): 67-76.
2.   Friesen, M.D. and L. Garren. 1982.   International
     Mycotoxin Check Sample Program Part I. Report on

the Performance of Participating Laboratories for the Analysis of Aflatoxins B1, B2, G1 and G2 in Raw Peanut Meal, De-oiled Peanut Meal and Yellow Corn Meal. J. Assoc. Off. Anal. Chem. 65:855-863.

3. Australian Standard 2580-1986. Chemical Analysis Interlaboratory Test Program for Determining Precision of Analytical Method(s). Guide to the Planning and Conduct.

4. ISO-5725 (1989). Precision of Test Methods – Determination of Repeatability and Reproducibility of Interlaboratory Tests.

International Survey on Dietary Fiber Definition, Analysis and Reference Materials

S.C. Lee[1], L. Prosky[2], and J. Tanner[2]
[1]KELLOGG COMPANY, BATTLE CREEK, MI 49016, USA
[2]US FOOD AND DRUG ADMINISTRATION, LAUREL, MD 20708, USA

Current definitions of dietary fiber reflect different opinions(1-3). An international survey has been conducted to solicit opinions from professionals in the dietary fiber field. This survey was designed to determine if a consensus could be reached on the definition of dietary fiber and associated methodologies. The survey also was designed to indicate what future directions for improving analytical methodology and reference materials would be considered appropriate as well as considering appropriate classification of non-enzymatically hydrolyzable oligosaccharides. The statistical results of this survey should assist international harmonization of acceptable definitions of dietary fiber and its components. The survey results will also provide insight for chemists seeking to improve analytical methods and reference materials to meet that definition. This preliminary report summarizes the 66 responses received so far. Complete results are expected to be published in the Journal of Association of Official Analytical Chemists (AOAC) International.

Survey Method :
Approximately 200 survey forms have been distributed to the professionals in the dietary fiber/carbohydrate area. The respondents consist of researchers in North America, Europe, Latin America, Asia and Australia.

1. What do you think is the most appropriate basis for the definition of dietary fiber? (Check only one.)

    a. From chemical perspectives only
       (i.e. analytical definition.)
    b. From both physiological and chemical
       perspectives.
    c. From physiological perspectives only.
    d. Others _____.

2. How would you define dietary fiber? (Check only one.)

   a. <u>Remnants of plant components</u> resistant to human alimentary enzymes. These include undigestible polysaccharides, lignin, undigestible protein and lipids, etc.
   b. <u>The sum of lignin and polysaccharides</u> that are not hydrolyzed by human alimentary enzymes, i.e. the sum of non-starch polysaccharides, resistant starch and lignin.
   c. <u>Polysaccharides</u> that are not hydrolyzed by human alimentary enzymes, i.e. the sum of non-starch polysaccharides.
   d. The sum of non-starch polysaccharides plus lignin.
   e. Plant cell wall components.
   f. Non-starch polysaccharides only.
   g. Others _____

3. Oligosaccharides that are not hydrolyzed by human alimentary enzymes should be included in the dietary fiber definition.

   a. Support (i.e. non-hydrolyzable oligosaccharides should be classified as dietary fiber).
   b. Not support (i.e. non-hydrolyzable oligosaccharides should be classified separately from dietary fiber).
   c. Others _____

4. It has been suggested that the term dietary fiber should be obsolete from scientific literature.

   a. Support (i.e. the term DF should be abolished).
   b. Not support (i.e. the term DF should be reserved).

5. The definition of soluble and insoluble dietary fiber should be based on:

   a. Physiological solubility of dietary fiber escaping human small intestine.
   b. Chemical solubility of dietary fiber under a defined chemical treatment condition.
      Example: DMSO-soluble DF.
               Hot buffered neutral detergent-soluble DF and others.
   c. Others _____

6. Based on your definition and the current available techniques, which method do you think is the most appropriate treatment prior to separating soluble and insoluble dietary fiber?

a. Hot DMSO (dimethylsulfoxide) solution and subsequent enzymatic treatments.
b. Autoclave (121°C, 1-1/2h), and subsequent enzymatic treatments.
c. Gelatinization at 90–100°C and enzymatic treatment.
d. Hot buffered neutral detergent treatment.
e. No treatment.
f. Others _____

7. Among the current techniques, which series of enzymes would you recommend the most appropriate to prepare dietary fiber residues? You may check up to two (2), with rank. Score 1 indicates the option you support the most, and score 2 for the second best option.

a. Termamyl, pullulanase and pancreatin. (Englyst method)
b. Termamyl and amyloglucosidase. (Theander method)
c. Termamyl, protease and amyloglucosidase. (AOAC method)
d. Termamyl, pepsin and pancreatin. (Asp et al)
e. Pepsin, pancreatin and amyloglucosidase. (Schweizer & Wursch)
f. Others _____

8. You are asked to design a dietary fiber method as a future method. Which enzymes would you choose to prepare dietary fiber residues and explain why.

9. Among the current available techniques, which type of method would you recommend as the most appropriate for dietary fiber labeling purposes? (Check only one.)

a. Enzymatic – colorimetric.
b. Enzymatic – GLC.
c. Enzymatic – HPLC.
d. Enzymatic – gravimetric.
e. Other (specify) _____

10. Based on current available techniques, which type of method would you recommend as the most appropriate for quality control research purposes, i.e. controlling or monitoring dietary fiber levels in raw ingredients and finished foods. (Check only one.)

a. Enzymatic – colorimetric.
b. Enzymatic – GLC.
c. Enzymatic – HPLC.
d. Enzymatic – gravimetric.
e. Other (name) _____

11. Based on current available techniques, which type of method would you recommend as the most appropriate for nutrition research purposes. (Check only one.)

a. Enzymatic - colorimetric.
b. Enzymatic - GLC.
c. Enzymatic - HPLC.
d. Enzymatic - gravimetric.
e. Other (name) _____

12. As a future method, which type of method would you recommend for dietary fiber labeling purposes?

a. Enzymatic - colorimetric.
b. Enzymatic - GLC.
c. Enzymatic - HPLC.
d. Enzymatic - gravimetric.
e. Other (name) _____

13. As a future method, which type of method would you recommend for quality control research purposes:

a. Enzymatic - colorimetric.
b. Enzymatic - GLC.
c. Enzymatic - HPLC.
d. Enzymatic - gravimetric.
e. Other (name) _____

14. As a future method, which type of method would you recommend for nutrition research purposes?

a. Enzymatic - colorimetric.
b. Enzymatic - GLC.
c. Enzymatic - HPLC.
d. Enzymatic - gravimetric.
e. Other (name) _____

15. For dietary fiber labeling purposes, which should be listed in the nutrition column labeling on food packages? (Check only one.)

a. Non-starch polysaccharides only.
b. Total, soluble and insoluble NSP.
c. Separate list of non-starch polysaccharides, (NSP) with optional soluble & insoluble breakdown resistant starch and lignin.
d. Total dietary fiber only.
e. Total, soluble, and insoluble dietary fiber.
f. Separate list of cellulose, soluble/insoluble hemicelluloses, soluble/insoluble beta-glucans, resistant starch, lignin and others.
g. As a most detailed list as possible, i.e. fractions listed in f. plus neutral sugar and uronic acid composition of each fraction.

16. For nutrition research purposes, which information should be available?

    a. Non-starch polysaccharides only.
    b. Total, soluble and insoluble NSP.
    c. Separate list of non-starch polysaccharides, (NSP) with optional soluble & insoluble breakdown resistant starch and lignin.
    d. Total dietary fiber only.
    e. Total, soluble, and insoluble dietary fiber.
    f. Separate list of cellulose, soluble/insoluble hemicelluloses, soluble/insoluble beta-glucans, resistant starch, lignin and others.
    g. As a most detailed list as possible, i.e. fractions listed in f. plus neutral sugar and uronic acid composition of each fraction.
    h. Others (Specify) _____

    _____

17. What should be done in the future to further improve the dietary fiber analysis methodology?

18. What do you use as an analytical standard(s) for quality control purposes? (ex. certified reference materials, in-house standard or others.)

19. Do you use certified standards available from NIST or reference materials from BCR in your laboratory?

20. Your main research interest is in: (Check only one.)

    a. Nutrition research.
    b. Analytical.
    c. Food Technology.
    d. Medical Research.
    e. Other (list) _____

21. Your affiliation is with: (Check only one.)

    a. University.
    b. Research Institute.
    c. Industry.
    d. Government/States.

22. Your geographical location is:

    a. North America.
    b. Europe.
    c. Pacific Rim - Australia and Asia
    d. Latin America.
    e. Africa.

23. Your country is:

_____

24. Your specialties are in:

Dietary Fiber.
General Carbohydrates.
Other Nutrients.
Other (specify) _____

Summary:
The preliminary survey results indicated that the researchers support the following.
1. The definition of dietary fiber should be based on both chemical and physiological perspectives.
2. Dietary fiber is defined as either the sum of lignin and polysaccharides that are not hydrolyzed by human alimentary enzymes(i.e. dietary fiber as the sum of resistant starch, nonstarch polysaccharides, and lignin) or the remnant of plant components resistant to human alimentary enzymes. The former has received more support than the latter.
3. The term dietary fiber should be preserved.
4. The oligosaccharides which escape the human small intestine are considered dietary fiber.
5. The definition of soluble and insoluble dietary fiber (SDF/IDF) should be based on either physiological solubility of dietary fiber escaping the human small intestine.
6. There is no consensus on which pretreatment is the best prior to separating SDF and IDF. However, gelatinization at 95–100 °C and following enzyme treatment has received more support than the other approaches.
7. Among the current techniques, the enzyme system which AOAC methods 985.29, 991.42 and 991.43 employ has been strongly supported as the most appropriate for the preparation of dietary fiber residues.
8. As a future approach, an enzymatic digestion system which could simulate a human gastric enzyme mixture (i.e. physiological enzyme approach) would be supported to prepare dietary fiber residues.
9. Among the currently available techniques, enzymatic- gravimetric methods are recommended as the most appropriate for both nutrition labeling and quality control purposes.
10. For nutrition research purposes, both enzymatic-gravimetric, GLC, and HPLC methods are supported as the most appropriate among currently available techniques.

11. As a future method, enzymatic–gravimetric methods are considered the most appropriate for both food labeling and quality control purposes.
12. Enzymatic–HPLC methods are considered as the most appropriate for future nutrition research purposes. Enzymatic–gravimetric methods also received support as the second best approach.
13. For nutrition labeling purposes, listing of TDF alone or TDF, SDF and IDF has received the most support, as opposed to non–starch polysaccharides.
14. For nutrition research purposes, researchers want to have the most detailed dietary fiber components list as possible.
15. For future improvements in analytical methodology, researchers have suggested the following:
    a. Add a fat digestion step.
    b. Improve HPLC–type methods for reliable characterization of DF components.
    c. Improve enzyme digestion steps to simulate the human physiological enzymatic system.
    d. Improve enzymatic–colorimetric methods for SDF and IDF quantification, and correlate these values to values obtained by more detailed methods(HPLC or GLC).
    e. Standardize sample preparation steps.
    f. More actively develop certified reference materials for dietary fiber analysis.
    g. Reach a consensus on definition of TDF, SDF and IDF.
    h. Solve the current problems associated with discrepancy of the values obtained by colorimetric and GLC approaches.
16. Only 20 % of the laboratories surveyed use an analytical standard for data quality control purposes. In–house standards, reference materials available from European Community Bureau of Reference and American Association of Cereal Chemists(AACC) check samples are being used as standards.

References:
1. Trowell, H. Ischemic heart disease and dietary fiber. Am.J.Clin.Nutr. 25;926, 1972.
2. Trowell, H.C., Southgate, D.A.T., Wolever, T.M.S., Leeds, A.R., Gassull, M.A., and Jenkins, D.J.A. Dietary fiber redefined. Lancet 1:967,1976.
3. The British Nutrition Foundation. Complex Carbohydrates in Foods. Chapman & Hall, London, 1990.

General Terms and their Definitions Concerning Standardization and Related Activities

ISO/IEC GUIDE 2:1991

1.6.1 international standardization: *Standardization* in which involvement is open to relevant *bodies* from all countries.

1.6.2 regional standardization: *Standardization* in which involvement is open to relevant *bodies* from countries from only one geographical, political or economic area of the world.

1.6.3 national standardization: *Standardization* that takes place at the level of one specific country.

3.1 normative document: Document that provides rules, guidelines or characteristics for activities or their results.

NOTES

1 The term *"normative document"* is a generic term that covers such documents as *standards, technical specifications, codes of practice* and *regulations*.

2 A document is to be understood as any medium with information recorded on or in it.

3 The terms for different kinds of *normative documents* are defined considering the document and its content as a single entity.

3.2   standard: Document, established by *consensus* and approved by a recognized *body*, that provides, for common and repeated use, rules, guidelines or characteristics for activities or their results, aimed at the achievement of the optimum degree of order in a given context.

NOTE — *Standards* should be based on the consolidated results of science, technology and experience, and aimed at the promotion of optimum community benefits.

3.2a)   prestandard: Document that is adopted provisionally by a *standardizing body* and made available to the public in order that the necessary experience may be gained from its *application* on which to base a *standard.*

3.3   technical specification: Document that prescribes technical *requirements* to be fulfilled by a product, process or service.

NOTES

1   A *technical specification* should indicate, whenever appropriate, the procedure(s) by means of which it may be determined whether the *requirements* given are fulfilled.

2   A *technical specification* may be a *standard*, a part of a *standard* or independent of a *standard*.

3.4   code of practice: Document that recommends practices or procedures for the design, manufacture, installation, maintenance or utilization of equipment, structures or products.

NOTE — A *code of practice* may be a *standard*, a part of a *standard* or independent of a *standard*.

3.5   regulation: Document providing binding legislative rules, that is adopted by an *authority*.

4.4 **standards body**: *Standardizing body* recognized at national, regional or international level, that has as a principal function, by virtue of its statutes, the preparation, approval or adoption of *standards* that are made available to the public.

NOTE — A *standards body* may also have other principal functions.

4.4.1 **national standards body**: *Standards body* recognized at the national level, that is eligible to be the national member of the corresponding *international* and *regional standards organizations*.

4.4.2 **regional standards organization**: *Standards organization* whose membership is open to the relevant national *body* from each country within one geographical, political or economic area only.

4.4.3 **international standards organization**: *Standards organization* whose membership is open to the relevant national *body* from every country.

4.5 **authority**: *Body* that has legal powers and rights.

NOTE — An *authority* can be regional, national or local.

4.5.1 **regulatory authority**: *Authority* that is responsible for preparing or adopting *regulations*.

4.5.2 **enforcement authority**: *Authority* that is responsible for enforcing *regulations*.

NOTE — The *enforcement authority* may or may not be the same as the *regulatory authority*.

5 Types of standards

NOTE — The following terms and definitions are not intended to provide a systematic classification or comprehensive list of possible types
of *standards*. They indicate some common types only. These are not
mutually exclusive; for instance, a particular *product standard* may
also be regarded as a *testing standard* if it provides *methods of test* for
characteristics of the product in question.

5.1 basic standard: *Standard* that has a wide-ranging
coverage or contains general *provisions* for one particular field.

NOTE — A *basic standard* may function as a *standard* for direct application or as a basis for other *standards*.

5.3 testing standard: *Standard* that is concerned with *test
methods*, sometimes supplemented with other *provisions*
related to *testing*, such as sampling, use of statistical methods,
sequence of tests.

5.4 product standard: *Standard* that specifies *requirements* to be fulfilled by a product or a group of products, to
establish its *fitness for purpose*.

NOTES

1 A *product standard* may include in addition to the *fitness for purpose requirements*, directly or by reference, aspects such as termin-
ology, sampling, *testing*, packaging and labelling and, sometimes, processing *requirements*.

2 A *product standard* can be either complete or not, according to
whether it specifies all or only a part of the necessary *requirements*. In
this respect one may differentiate between *standards* such as dimen-
sional, material, and technical delivery *standards*.

5.5 process standard: *Standard* that specifies *requirements* to be fulfilled by a process, to establish its *fitness for
purpose*.

12 Testing

NOTE — Definitions of terms relevant to *testing* but relating to measurement standards can be found in the *International vocabulary of basic and general terms in metrology (VIM)*.

12.1   test: Technical operation that consists of the determination of one or more characteristics of a given product, process or service according to a specified procedure.

12.1.1   testing: Action of carrying out one or more *tests*.

12.2   test method: Specified technical procedure for performing a *test*.

12.3   test report: Document that presents *test* results and other information relevant to a *test*.

12.4   testing laboratory: Laboratory that performs *tests*.

NOTE — The term *"testing laboratory"* can be used in the sense of a legal entity, a technical entity or both.

12.5   interlaboratory test comparisons: Organization, performance and evaluation of *tests* on the same or similar items or materials by two or more laboratories in accordance with predetermined conditions.

12.6   (laboratory) proficiency testing: Determination of laboratory testing performance by means of *interlaboratory test comparisons*.

13   Conformity and related general concepts

13.1   conformity: Fulfilment by a product, process or service of specified *requirements*.

13.2   third party: Person or *body* that is recognized as being independent of the parties involved, as concerns the issue in question.

NOTE — Parties involved are usually supplier ("first party") and purchaser ("second party") interests.

13.3   evaluation for conformity: Systematic examination of the extent to which a product, process or service fulfils specified *requirements*.

13.3.1   inspection: *Evaluation for conformity* by measuring, observing, *testing* or gauging the relevant characteristics.

13.3.2   conformity testing: *Evaluation for conformity* by means of *testing*.

13.4   verification of conformity: Confirmation, by examination of evidence, that a product, process or service fulfils specified *requirements*.

13.5   assurance of conformity: Procedure resulting in a statement giving confidence that a product, process or service fulfils specified *requirements.*

NOTE — For a product, the statement may be in the form of a document, a label or other equivalent means. It may also be printed in or applied on a communication, a catalogue, an invoice, a user instructions manual, etc. relating to the product.

13.5.1 supplier's declaration: Procedure by which a supplier gives written assurance that a product, process or service conforms to specified *requirements.*

NOTE — In order to avoid any confusion, the expression "self-certification" should not be used.

13.5.2 certification: Procedure by which a *third party* gives written assurance that a product, process or service conforms to specified *requirements.*

13.6 registration: Procedure by which a *body* indicates relevant characteristics of a product, process or service, or particulars of a *body* or person, in an appropriate, publicly available list.

13.7 accreditation: Procedure by which an authoritative *body* gives formal recognition that a *body* or person is competent to carry out specific tasks.

14 Certification activities

14.1 certification system: System that has its own rules of procedure and management for carrying out *certification of conformity.*

NOTES

1 *Certification systems* may be operated at, for example, national, regional or international level.

2 The central *body* that conducts and administers a *certification system* may decentralize its activities and rights to certify *conformity.*

14.2 certification scheme: *Certification system* as related to specified products, processes or services to which the same particular *standards* and rules, and the same procedure, apply.

NOTE — The term "certification programme" is used in some countries to cover the same concept as *"certification scheme".*

14.3   certification body : *Body* that conducts *certification of conformity*.

NOTE — A *certification body* may operate its own testing and inspection activities or oversee these activities carried out on its behalf by other *bodies*.

14.4   inspection body (for certification) : *Body* that performs inspection services on behalf of a *certification body*.

14.5   licence (for certification) : Document, issued under the rules of a *certification system*, by which a *certification body* grants to a person or *body* the right to use *certificates* or *marks of conformity* for its products, processes or services in accordance with the rules of the relevant *certification scheme*.

14.6   applicant (for certification) : Person or *body* that seeks to obtain a *licence* from a *certification body*.

14.7   licensee (for certification) : Person or *body* to which a *certification body* has granted a *licence*.

14.8   certificate of conformity : Document issued under the rules of a *certification system*, indicating that adequate confidence is provided that a duly identified product, process or service is in *conformity* with a specific *standard* or other *normative document*.

16   Accreditation of testing laboratories

NOTE — For ease of reading, the term "laboratory" has been used for *"testing laboratory"* in the terms defined below.

16.1 **(laboratory) accreditation**: Formal recognition that a *testing laboratory* is competent to carry out specific *tests* or specific types of *tests*.

NOTE — The term *"laboratory accreditation"* may cover the recognition of both the technical competence and the impartiality of a *testing laboratory* or only its technical competence. *Accreditation* is normally awarded following successful laboratory assessment and is followed by appropriate surveillance.

16.2 **(laboratory) accreditation system**: System that has its own rules of procedure and management for carrying out *laboratory accreditation*.

16.3 **(laboratory) accreditation body**: *Body* that conducts and administers a *laboratory accreditation system* and grants *accreditation*.

NOTE — An *accreditation body* may wish to delegate fully or partially the assessment of a *testing laboratory* to another competent *body* (assessment agency). Whilst it is recognized that this may be a practical solution to extending recognition of *testing laboratories*, it is essential that such assessment be equivalent to that applied by the *accreditation body* and that the *accreditation body* take full responsibility for such extended *accreditation*.

16.4 **accredited laboratory**: *Testing laboratory* to which *accreditation* has been granted.

16.5 **(laboratory) accreditation criteria**: Set of *requirements* that is used by an *accreditation body*, to be fulfilled by a *testing laboratory* in order to be accredited.

16.6   laboratory assessment: Examination of a *testing laboratory* to evaluate its compliance with specific *laboratory accreditation criteria*.

16.7   laboratory assessor: Person who carries out some or all functions related to *laboratory assessment*.

16.8   accredited laboratory test report: *Test report* that includes a statement by the *testing laboratory* that it is accredited for the *test* reported and that the *test* has been performed in accordance with the conditions prescribed by the *accreditation body*.

16.9   approved signatory (of an accredited laboratory): Person who is recognized by an *accreditation body* as competent to sign *accredited laboratory test reports*.

Selective List of International Standards

Statistical methods

ISO 2602:1980 Statistical interpretation of test results-- Estimation of the mean-- Confidence interval
Ed. 2 5 p. Code C   TC 69 HB 3

ISO 2854:1976 Statistical interpretation of data-- Techniques of estimation and tests relating to means and variances

Ed. 1 46 p. Code T   TC 69 HB 3

DIS 2859-0 Sampling procedures for inspection by attributes-- Part 0:Introduction to the ISO 2859 attribute sampling system (Revision of ISO 2859:1974 and of Addendum 1:1977)

Ed. 1 96 p. Code TC 69

ISO 2859-1:1989 Sampling procedures for inspection by attributes-- Part 1:Sampling plans indexed by acceptable quality level (AQL) for lot-by-lot inspection

Ed. 1 67 p. Code V   TC 69 HB 3

ISO 2859-2:1985 Sampling procedures for inspection by attributes-- Part 2:Sampling plans indexed by limiting quality (LQ) for isolated lot inspection

Ed. 1 21 p. Code L   TC 69 HB 3

ISO 2859-3:1991 Sampling procedures for inspection by attributes-- Part 3:Skip-lot sampling procedures

Ed. 1 16 p. Code H   TC 69

ISO 3207:1975 Statistical interpretation of data-- Determination of a statistical tolerance interval

Ed. 1 15 p. Code H   TC 69 HB 3

Addendum 1:1978 to ISO 3207:1975
Ed. 1 3 p. Code B   TC 69 HB 3

ISO 3301:1975 Statistical interpretation of data-- Comparison of two means in the case of paired observations

Ed. 1 6 p. Code C   TC 69 HB 3

ISO 3494:1976	Statistical interpretation of data-- Power of tests relating to means and variances *Ed. 1 44 p. Code S TC 69 HB 3*
ISO 3534:1977	Statistics-- Vocabulary and symbols Bilingual edition *Ed. 1 47 p. Code T TC 69 HB 3*
DIS 3534-1	Statistics-- Vocabulary and symbols-- Part 1: Probability and general statistical terms (Revision of ISO 3534:1977) *Ed. 1 0 p. Code W TC 69*
DIS 3534-2	Statistics-- Vocabulary and symbols-- Part 2:Statistical quality control (Revision, in part, of ISO 3534-1977) Bilingual edition *Ed. 1 71 p. Code P TC 69*
ISO 3534-3:1985	Statistics-- Vocabulary and symbols-- Part 3:Design of experiments Bilingual edition *Ed. 1 33 p. Code Q TC 69 HB 3*
ISO 3951:1989	Sampling procedures and charts for inspection by variables for percent nonconforming *Ed. 2 107 p. Code XA TC 69 HB 3*
ISO 5725:1986	Precision of test methods-- Determination of repeatability and reproducibility for a standard test method by inter-laboratory tests *Ed. 2 49 p. Code T TC 69 HB 3*
DIS 5725-1	Accuracy (trueness and precision) of measurement methods and results-- Part 1:General principles and definitions (Revision, in parts, ofISO 5725:1986) *Ed. 1 0 p. Code TC 69*
DIS 5725-2	Accuracy (trueness and precision) of measurement methods and results-- Part 2: A basic method for the determination of repeatability and reproducibility of a standard measurement method (Revision, in parts, of ISO 5725:1986) *Ed. 1 0 p. Code TC 69*
DIS 5725-3	Accuracy (trueness and precision) of measurement methods and results-- Part 3: Intermediate measures of the precision of a measurement method *Ed. 1 46 p. Code TC 69*

DIS 5725-4	Accuracy (trueness and precision) of measurement methods and results-- Part 4:Basic methods for estimating the trueness of a test method (Revision, in parts, of ISO 5725:1986)
	Ed. 1 0 p. Code TC 69
DIS 5725-6	Accuracy (trueness and precision) of measurement methods and results-- Part 6: Practical applications (Revision, in parts, of ISO 5725:1986)
	Ed. 1 76 p. Code TC 69
DIS 7870	Control charts-- General guide and introduction
	Ed. 1 0 p. Code D TC 69
DIS 7873	Control charts for arithmetic average with warning limits
	Ed. 1 0 p. Code G TC 69
DIS 7966	Acceptance control charts
	Ed. 1 0 p. Code K TC 69
ISO 8258:1991	Shewhart control charts
	Ed. 1 29 p. Code N TC 69
ISO 8422:1991	Sequential sampling plans for inspection by attributes
	Ed. 1 45 p. Code S TC 69
ISO 8423:1991	Sequential sampling plans for inspection by variables for percent nonconforming (known standard deviation)
	Ed. 1 39 p. Code R TC 69
ISO 8595:1989	Interpretation of statistical data-- Estimation of a median
	Ed. 1 3 p. Code B TC 69
DIS 11453	Statistical interpretation of data-- Tests and confidence intervals relating to proportions
	Ed. 1 73 p. Code TC 69

Handbook 3 — Statistical methods

Statistical processing and interpretation of test and inspection results. Includes the two basic tools used in sampling throughout the world — sampling by attributes and by variables — several standards relating to the use of data which are averaged, an important standard on the precision of test methods, and a vocabulary.

Price group X. 456 pages. (3rd ed. 1989)
ISBN 92-67-10153-6

Quality assurance

ISO 8402:1986	Quality-- Vocabulary Trilingual edition *Ed. 1 12 p. Code F TC 176*
DIS 8402	Quality management and quality assurance-- Vocabulary (Revision of ISO 8402:1986) Trilingual edition *Ed. 2 0 p. Code TC 176*
ISO 9000:1987	Quality management and quality assurance standards-- Guidelines for selection and use *Ed. 1 6 p. Code C TC 176*
DIS 9000-2	Quality management and quality assurance standards-- Part 2:Generic guidelines for the application of ISO 9001, ISO 9002 and ISO 9003 *Ed. 1 35 p. Code TC 176*
ISO 9000-3:1991	Quality management and quality assurance standards-- Part 3:Guidelines for the application of ISO 9001 to the development, supply and maintenance of software *Ed. 1 15 p. Code H TC 176*
DIS 9000-4	Quality management and quality assurance standards-- Part 4:Application for dependability management *Ed. 1 9 p. Code TC 176*
ISO 9001:1987	Quality systems-- Model for quality assurance in design/development, production, installation and servicing *Ed. 1 7 p. Code D TC 176*
ISO 9002:1987	Quality systems-- Model for quality assurance in production and installation *Ed. 1 6 p. Code C TC 176*
ISO 9003:1987	Quality systems-- Model for quality assurance in final inspection and test *Ed. 1 2 p. Code A TC 176*

ISO 9004:1987	Quality management and quality system elements-- Guidelines *Ed. 1 16 p. Code H TC 176*
ISO 9004-2:1991	Quality management and quality system elements-- Part 2:Guidelines for services *Ed. 1 18 p. Code J TC 176*
DIS 9004-3	Quality management and quality system elements-- Part 3:Guidelines for processed materials *Ed. 2 41 p. Code TC 176*
DIS 9004-4	Quality management and quality system elements-- Part 4:Guidelines for quality improvement *Ed. 2 45 p. Code TC 176*
ISO 10011-1:1990	Guidelines for auditing quality systems-- Part 1:Auditing *Ed. 1 7 p. Code D TC 176*
ISO 10011-2:1991	Guidelines for auditing quality systems-- Part 2:Qualification criteria for quality systems auditors *Ed. 1 5 p. Code C TC 176*
ISO 10011-3:1991	Guidelines for auditing quality systems-- Part 3:Management of audit programmes *Ed. 1 3 p. Code B TC 176*
ISO 10012-1:1992	Quality assurance requirements for measuring equipment-- Part 1:Metrological confirmation system for measuring equipment *Ed. 1 14 p. Code G TC 176*

ISO 9000 Compendium — International Standards for quality Management

This compendium contains both International Standards and draft International Standards, known as the ISO 9000 series. The book comes together with a paper "Vision 2000 — A strategy for International Standards' implementation in the quality arena during the 1990s".

1991, 176 pages   *ISBN 92-67-10165-X*

Guides

In addition to International Standards, ISO publishes guides covering subjects related to international standardization.

ISO/IEC Guide 2:1991
General terms and their definitions concerning standardization and related activities
Trilingual edition
Ed. 6   60 p.   Code U

ISO/IEC Guide 7:1982
Requirements for standards suitable for product certification
Ed. 1   4 p.   Code B

ISO/IEC Guide 16:1978
Code of principles on third party certification systems and related standards
Ed. 1   2 p.   Code A

ISO/IEC Guide 22:1982
Information on manufacturer's declaration of conformity with standards or other technical specifications
Ed. 1   4 p.   Code B

ISO/IEC Guide 23:1982
Methods of indicating conformity with standards for third-party certification systems
Ed. 1   4 p.   Code B

ISO/IEC Guide 25:1990
General requirements for the competence of calibration and testing laboratories
Ed. 3   7 p.   Code D

ISO Guide 27:1983
Guidelines for corrective action to be taken by a certification body in the event of misuse of its mark of conformity
Ed. 1   5 p.   Code C

ISO/IEC Guide 28:1982
General rules for a model third-party certification system for products
Ed. 1   16 p.   Code H

ISO Guide 30:1992
Terms and definitions used in connection with reference materials
Bilingual edition
Ed. 2   8 p.   Code D

ISO Guide 31:1981
Contents of certificates of reference materials
Ed. 1   8 p.   Code D

ISO Guide 33:1989
Uses of cerfified reference materials
Ed. 1   12 p.   Code F

ISO Guide 35:1989
Certification of reference materials — General and statistical principles
Ed. 2   32 p.   Code P

ISO/IEC Guide 39:1988
General requirements for the acceptance of inspection bodies
Ed. 2   8 p.   Code D

ISO/IEC Guide 40:1983
General requirements for the acceptance of certification bodies
Ed. 1   3 p.   Code B

ISO/IEC Guide 42:1984
Guidelines for a step-by-step approach to an international certification system
Ed. 1   6 p.   Code C

ISO/IEC Guide 43:1984
Development and operation of laboratory proficiency testing
Ed. 1   6 p.   Code C

ISO/IEC Guide 44:1985
General rules for ISO or IEC international third-party
certification schemes for products
Ed. 1   13 p.   Code G

ISO/IEC Guide 48:1986
Guidelines for third-party assessment and registration of a
supplier's Quality System
Ed. 1   9 p.   Code E

ISO/IEC Guide 53:1988
An approach to the utilization of a supplier's quality system
in third party product certification
Ed. 1   13 p.   Code G

ISO/IEC Guide 56:1989
An approach to the review by a certification body of its own
internal quality system
Ed. 1   4 p.   Code B

ISO/IEC Guide 57:1991
Guidelines for the presentation of inspection results
Ed. 1   3 p.   Code B

ISO/IEC Guide 58:1993
Calibration and testing laboratory accreditation
systems — General requirements for operation and
recognition
Ed. 1   6 p.   Code C

Addresses of Authors

A. R. Aguinaldo
Food and Nutrition Research Institute Department
of Science and Technology

Pedro Gil St. cor. Taft Ave
Ermita 1000, Manila, Philippines

Richard Albert
U.S. Food and Drug Administration

Washington
DC 20204, USA

S. M. Anderson
Kellogg Company, Science and Technology Center

235 Porter Street, Battle Creek
MI 49017-6210, USA

J. G. Ardena
Food and Nutrition Research Institute Department
of Science and Technology

Pedro Gil St. cor. Taft Ave
Ermita 1000, Manila, Philippines

James A. Ault
ABC Laboratories

Inc., PO Box 1097
Columbia, MO 65205, USA

E. M. Avena
Food and Nutrition Research Institute Department
of Science and Technology

Pedro Gil St. cor. Taft Ave
Ermita 1000, Manila, Philippines

Trean Korbelak Blumenthal, M. S.
Director of Corporate Quality Assurance

Libra Laboratories Inc.
16 Pearl Street, Metuchen
NJ 08840, USA

C. Bruhn
Instrumental Analysis Department, Pharmacy
Faculty. University of Conceptión

Casilla 237
Conceptión, CHILE

A. Y. Cantillo
National Oceanic and Atmospheric Administration

NOS/ORCA N/ORCA 21
Rockville, MD 20852, USA

E. M. Castillo
Food and Nutrition Research Institute Department
of Science and Technology

Pedro Gil St. cor. Taft Ave
Ermita 1000, Manila, Philippines

Robert W. Dabeka
Food Research Division, Bureau of Chemical Safety
Food Directorate, Health Protection Branch

Health and Welfare Canada
Ottawa, Ontario, Canada K1A OL2

Ch.-G. de Boroviczény
Institut für Standardisierung und Dokumentation im
medizinischen Laboratorium (INSTAND) e. V.

Haslacherstrasse 51
D-7800 Freiburg, Germany

D. Delgado
Chemical Institute,
Catholic University of Valparaiso   Casilla 4059
                                                     Valparaiso, CHILE

I. Do Gregori
Chemical Institute,
Catholic University of Valparaiso   Casilla 4059
                                                     Valparaiso, CHILE

N. Gras
Neutron Activation Analysis Laboratory,
Chilean Energy Commission for Nuclear Energy   Casilla 188 D
                                                     Santiago, CHILE

A. H. Havelaar
Laboratory for Water and Food Microbiology   P.O. Box 1
                                                     3720 BA Bilthoven, NL

Stephen Hayward
Food Research Division, Bureau of Chemical Safety
Food Directorate, Health Protection Branch   Health and Welfare Canada
                                                     Ottawa, Ontario, Canada K1A OL2

S. H. Heisterkamp
Centre of Mathematical Methods, National Institute of
Public Health and Environmental Protection RIVM P.O. Box 1
                                                     3720 AL Bilthoven, NL

David N. Holcomb
Silliker Laboratories   1304 Halsted Street
                                                     Chicago Heights, IL 60411, USA

William Horwitz
U. S. Food and Drug Administration   Washington
                                                     DC 20204, USA

Bernard King
Laboratory of the Government Chemist   Queens Road, Teddington
                                                     Middlesex TW11 OLY, U.K.

Eugene J. Klesta
Chemical Waste Management Inc.   Alsip
                                                     IL 60658, USA

G. G. Lauenstein
National Oceanic and Atmospheric Administration NOS/ORCA N/ORCA 21
                                                     Rockville, MD 20852, USA

S.C. Lee
Kellogg Company   Battle Creek
                                                     MI 49016, USA

A. A. Liabastre
American Association for Laboratory Accreditation 656 Quince Orchard Road
                                                     Gaithersburg MD 20878, USA

J. W. Locke
American Association for Laboratory Accreditation 656 Quince Orchard Road
                                                     Gaithersburg MD 20878, USA

A. V. Lontoc
Food and Nutrition Research Institute Department
of Science and Technology   Pedro Gil St. cor. Taft Ave
                                               Ermita 1000, Manila, Philippines

P. M. Matibag
Food and Nutrition Research Institute Department
of Science and Technology   Pedro Gil St. cor. Taft Ave
                                               Ermita 1000, Manila, Philippines

Takashi Miyazu
Department of Management Engineering   The Nishi-Tokyo University
                                               Uenohara-machi Yamanashi-ken
                                               JAPAN 409-01

Alfredo M. Montes Niño
Veterinarian International Consultant and
Food Production   Buenos Aires
                                               Argentina

K. A. Mooijmann
Foundation for the Advancement of Public Health
and Environmental Protection (SVM)   P.O. Box 457
                                               3720 BA Bilthoven, NL

L. Muñoz
Neutron Activation Analysis Laboratory,
Chilean Energy Commission for Nuclear Energy   Casilla 188 D
                                               Santiago, CHILE

G. Navarrete
Intrumental Analysis Department, Pharmacy
Faculty. University of Conceptión   Casilla 237
                                               Conceptión, CHILE

J. Ngeh-Ngwainbi
Kellogg Company, Science and Technology Center 235 Porter Street, Battle Creek
                                               MI 49017-6210, USA

L. Paksy
Metalcontrol Kft Miskolc   P.O.B. 557, Miskolc
                                               H-3510, Hungary

M. Parkany
ISO Central Secretariat   Geneva 20, Case Postale 56
                                               CH-1211, Switzerland

R. M. Parris
National Institute of Standards and Technology
Organic Analytical Research Division   Gaithersburg
                                               MD 20899, USA

H. Pinochet
Chemical Institute,
Catholic University of Valparaiso   Casilla 4059
                                               Valparaiso, CHILE

T. R. Portugal
Food and Nutrition Research Institute Department
of Science and Technology   Pedro Gil St. cor. Taft Ave
                                                   Ermita 1000, Manila, Philippines

L. Prosky
U. S. Food and Drug Administration   Laurel
                                                   Maryland 20708, USA

Oscar D. Rampini
International Consultant on Food Analysiss   Buenos Aires
                                                   Argentina

R. S. Sagum
Food and Nutrition Research Institute Department
of Science and Technology   Pedro Gil St. cor. Taft Ave
                                                   Ermita 1000, Manila, Philippines

M. J. Seghatchian
Quality Department
North London Blood Transfusion Centre (NLBTC) London
                                                   England

J. F. A. Stivala
Quality Department
North London Blood Transfusion Centre (NLBTC) London
                                                   England

N. G. W. M. van Strijp-Lockefeer
Foundation for the Advancement of Public Health
and Environmental Protection (SVM)   P.O. Box 457
                                                   3720 AL, NL

J. Tanner
U. S. Food and Drug Administration   Laurel
                                                   Maryland 20708, USA

Hisashi Yamamoto
Department of Management Engineering   The Nishi-Tokyo University
                                                   Uenohara-machi Yamanashi-ken
                                                   JAPAN 409-01

Subject Index

A

A2LA (American Association for Laboratory Accreditation), 4, 5, 17
AACC (American Association of Cereal Chemists), 164
AAS (Atomic Absorption Spectroscopy), 4, 68, 144, 146
ABC Laboratories, Inc., 53
Acceptance criterion, 62, 65, 66
Accountability, 54
Accreditation, 4, 8, 18
 attendant, 4
 laboratories, 8, 128, 129
 organizations, 9
 systems, 8
Accuracy, 22, 26, 18, 37, 46, 66, 82, 108, 151, 152, 154
 of information, 102
 of observed values, 28
 within laboratory, 151
Aflatoxin, 89
AIHA (American Industrial Hygiene Association), 4, 18
Analytical
 chemistry, 56
 methodology, 158
 standard, 162
Annual appraisal, 57
ANOVA (Analysis of variance)
 1-way, 83
 2-way, 83
AOAC (Association of Official Analytical Chemists),International, 81, 89, 133
AOCS (American Oil Chemists' Society), 81
Archive, 16
 facility, 54
ASEANFOODS (ASEAN FOOD Data Network), 151
ASTM (American Society for Testing and Materials), 5, 6, 7, 15
Audit
 procedures, 54, 134
 trail, 54, 55
Automated cell counters, 103
Average value, 28

B

Background noise, 128
BCR (Community Bureau of Reference), 146, 162, 163
Benthic Surveillance Project, 37, 44
Bias, 23, 28, 82
 analyst, 82
 individual laboratory, 82
 laboratory, 65
 matrix, 82